Neuroanatomy of Language Regions
of the Human Brain

Neuroanatomy of Language Regions of the Human Brain

MICHAEL PETRIDES

Montreal Neurological Institute
McGill University
Montreal, Quebec
Canada

AMSTERDAM•BOSTON•HEIDELBERG•LONDON•NEW YORK•OXFORD
PARIS•SAN DIEGO•SAN FRANCISCO•SINGAPORE•SYDNEY•TOKYO
Academic Press is an imprint of Elsevier

Academic Press is an imprint of Elsevier
32 Jamestown Road, London NW1 7BY, UK
225 Wyman Street, Waltham, MA 02451, USA
525 B Street, Suite 1800, San Diego, CA 92101-4495, USA

First edition

British Library Cataloguing-in-Publication Data
A catalogue record for this book is available from the British Library

Library of Congress Cataloging-in-Publication Data
A catalog record for this book is available from the Library of Congress

ISBN : 978-0-12-405514-8

For information on all Academic Press publications
visit our website at www.store.elsevier.com

To my sister Soula,

very talented, caring, and devoted to everyone that is in her life and to everything she chooses to do.

Contents

Preface

There are many books on the aphasic disorders and the theoretical arguments regarding the functional role of the various core language areas (e.g., Caplan, 1992; Caplan and Hildebrandt, 1998; Nadeau, 2000; Hillis, 2002; Jackendoff, 2002; Pulvermüller, 2003; Gullberg and Indefrey, 2006; Ingram, 2007; Denes, 2011; Faust, 2012). The voluminous data now available from functional neuroimaging on the neural bases of cognitive processing, including language, has also been reviewed in several books (e.g., Pulvermüller, 2003; Cabeza and Kingstone, 2006) and articles (e.g., Démonet et al., 2005; Price, 2000, 2010). The present atlas does not intend to review the aphasic syndromes or the functional neuroimaging literature. It is strictly focused on the anatomical characteristics of the core language regions of the cerebral cortex, namely their morphology, cytoarchitecture, and connections.

The present atlas has been written at the encouragement of several colleagues whose research is focused on the neural bases of language. They pointed out that sophisticated linguistic and psychological dissection of language processes in neurolinguistic and functional neuroimaging studies is frequently accompanied by rather unsophisticated analysis of the anatomical correlates of these processes. Even a cursory examination of many functional neuroimaging studies purporting to examine the role of Broca's region in language illustrates the problem. Any activity related to language processing that appears within or near the ventral part of the precentral gyrus and the inferior frontal gyrus is declared to be Broca's area activation. But at least six distinct cytoarchitectonic areas are found in this general part of the cerebral hemisphere, not including the nearby frontal opercular areas: the orofacial part of the primary motor cortex (area 4), two ventral premotor areas (areas 6VC and 6VR), and three distinct parts of the inferior frontal gyrus (area 44 on the pars opercularis, area 45 on the pars triangularis, and area 47/12 on the pars orbitalis). All these cytoarchitectonic areas exhibit distinct cellular structure and distinct connectivity patterns. Thus, declaring any activity change in the ventrolateral frontal lobe to be in Broca's area

does not advance our understanding of the relative contribution of these distinct ventrolateral frontal areas to language processing, and may even hinder it by constructing functional models that fuse distinct aspects of processing which depend on different parts of this large and heterogeneous region. The same can be said of the parietal and temporal language regions where, again, the differences between multiple and distinct areas are ignored in the search for simple statements about the role of the parietal or temporal cortex in language.

The present atlas aims to be a useful guide to the anatomy of the core language regions of the cerebral cortex for the neurolinguist, aphasiologist, neuropsychologist, neurologist, speech therapist, and any researcher who uses structural or functional neuroimaging to explore the neural bases of language. It aims to present in a direct and comprehensible manner the available anatomical information on the core language system: the sulcal and gyral morphology of the core language regions, the cytoarchitecture of the cortical areas that are found in these regions, and the connectivity of these areas as studied by classical dissection methods, and more recently by diffusion magnetic resonance imaging (MRI) and resting state connectivity. The study of the details of the connections of the language areas of the cortex has been greatly aided by the discovery of cytoarchitectonically homologous regions in the macaque monkey brain. Although macaque monkeys do not possess language in the sense that humans do, study of the connections of the homologous areas provides critical information about the evolution of language and, more practically, precise information about the anatomical connections of these areas that cannot be achieved in the human brain, even with the recent developments of diffusion MRI and resting state connectivity. Indeed, as will become clear in the section on the connectivity of language areas, anatomical studies in the macaque monkey provide the key to deciphering the complex and controversial data obtained with the available methods in the human brain because the experimental tract tracing methods, which can be applied only on nonhuman primates, permit the precise definition of the origins, course, and terminations of particular pathways. The macaque monkey data are, unfortunately, relatively inaccessible to the language researcher because of major differences in the terminology of cytoarchitectonic areas used in the monkey literature in comparison with the human literature. It is hoped that the present atlas will help the reader navigate the complexities of the currently available data on the connectivity of the language areas of the cortex.

Acknowledgments

I am grateful to my technical assistants who helped put this atlas together. I am most indebted to Rhea Pavan who coordinated the production of this atlas in Adobe InDesign and prepared many of the illustrations. Jennifer Novek was responsible for the photography and labeling of the cytoarchitectonic areas and many of the anatomical illustrations and Trisanna Sprung-Much for the labeling of the MRI sections and the 3D reconstruction of the MRI volume. I would also like to thank Callah Boomhaur who worked with me in the early stages of the project and prepared some of the illustrations, and Claude Lepage for his technical assistance with regard to the 3D MRI reconstruction. I acknowledge support from the Centre of Excellence in Commercialization and Research, the Canadian Institutes of Health Research, and the Natural Sciences and Engineering Research Councils of Canada.

Historical Background

_____

Historical Background

The aim of this section of the atlas is to provide some historical background on the classical research that led to the definition of the core language regions of the brain. There is no doubt that virtually all cortical areas can contribute to language processing, depending on the particular requirements of the moment. However, this does not mean that all cortical areas are core language regions. A simple example will make the point. The mid-dorsolateral prefrontal cortex is central to the monitoring of information in working memory, verbal and nonverbal (e.g., Petrides et al., 1993; Petrides, 2013). Naturally, this cortical region is more or less engaged in verbal discourse or reading, depending on the degree to which monitoring of the information in working memory is necessary. This does not, however, make the mid-dorsolateral prefrontal region a core language area. Neither language production nor comprehension will be fundamentally impaired by damage to this region of the brain. Thus, core language regions are those whose contribution is central to language processing and their damage will lead to basic impairments in organizing and producing linguistic output, such as speaking and writing, or in comprehending such output, as in auditory comprehension and reading.

The term "core language regions" does not imply that these areas are uniquely, or even primarily, involved in language processing or that they are unique to the human brain. The same cytoarchitectonic areas exist in the nondominant hemisphere of the cortex where they are less essential to language and participate in non-linguistic processing. There is also increasing evidence that the cytoarchitectonic homologues of the so-called language areas exist in nonhuman primates (Petrides and Pandya, 1994, 2002; Petrides et al., 2005). Clearly, they are not used in language processing sensu stricto, although we can assume that their fundamental neural computations were essential to support language as it began evolving in the human brain. For further discussion of this issue, see last section on the connectivity of the language areas.

Broca's Region and Wernicke's Region: Historical Considerations

The scientific exploration of the brain regions that are critically involved in language processing started with the influential observations made by Pierre Paul Broca in Paris in the latter part of the 19th century. In 1861, Broca presented the cases of the two now famous patients who exhibited severe speech production impairments after damage to the posterior part of the ventrolateral frontal lobe in the left hemisphere (Broca, 1861a, b, c). The first case, Leborgne, was a 51 year old man who had lost his ability to speak several years earlier. When attempting to speak spontaneously or in response to questions, he managed to produce only the syllable "tan". At autopsy, a large lesion was noted in the posterior part of the third frontal convolution (inferior frontal gyrus) of the left hemisphere (Fig. 1). Broca considered this finding to be consistent with the claim, intensely discussed in Paris at the time, that the frontal lobe is critical for language and presented this case to the Anthropological Society (Broca, 1861a) and the Anatomical Society of Paris (Broca, 1861b).

FIGURE 1 Photographs of the lateral surface of the left hemisphere of the brains of Leborgne and Lelong to show the lesions in the inferior frontal gyrus. In Leborgne, the lesion extends beyond the inferior frontal gyrus and involves the nearby superior temporal region and the adjacent ventral anterior parietal region. Inspection of the expanded region (inset) allows one to appreciate the depth of the lesion, implying damage to the insula and adjacent basal ganglia. (Dronkers et al., Paul Broca's historic cases; high resolution MR imaging of the brains of Leborgne and Lelong, Brain, 2007, 130(5), pp. 1432-1441, by permission of Oxford University Press.)

Pierre Paul Broca
(1824 - 1880)

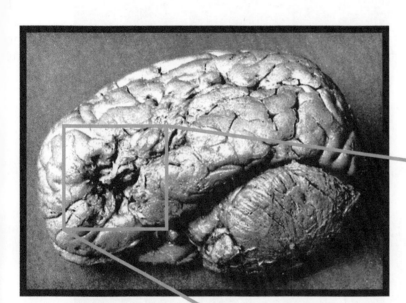

Leborgne

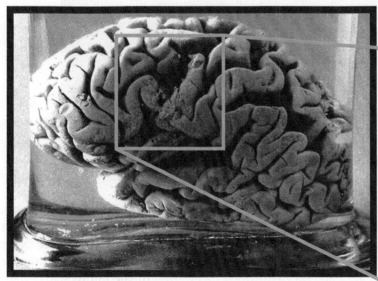

Lelong

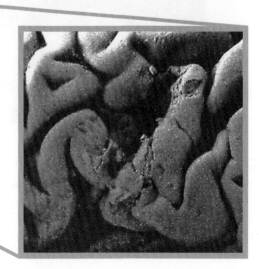

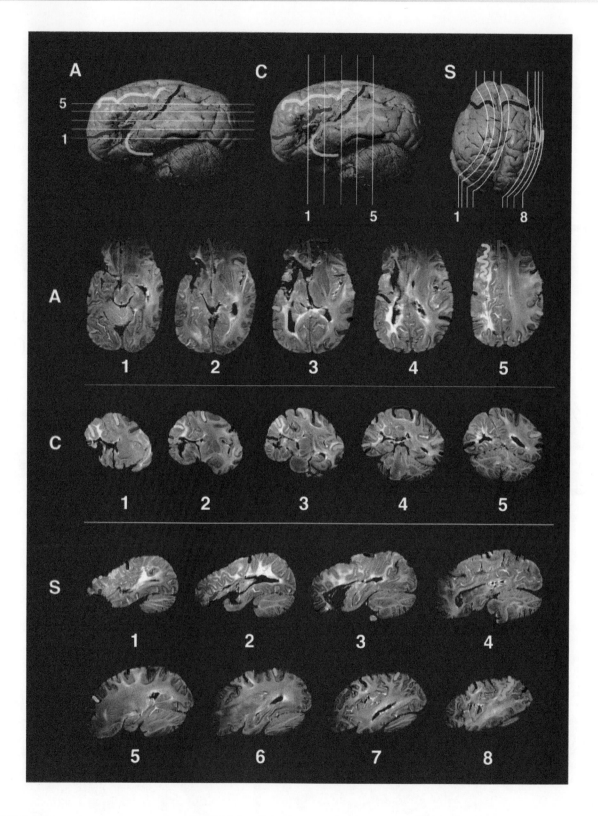

FIGURE 2 Magnetic resonance imaging (MRI) of the brain of Leborgne with axial (A), coronal (C) and sagittal (S) sections through the brain. The left hemisphere is on the left side in the axial and coronal sections. Colors mark the interhemispheric/longitudinal fissure (orange), central sulcus (dark blue), lateral fissure (aqua), inferior frontal sulcus (red), superior frontal sulcus (yellow), frontomarginal sulcus (pink), superior temporal sulcus (light green) and inferior temporal sulcus (brown). (Dronkers et al., Paul Broca's historic cases; high resolution MR imaging of the brains of Leborgne and Lelong, Brain, 2007, 130(5), pp. 1432-1441, by permission of Oxford University Press.)

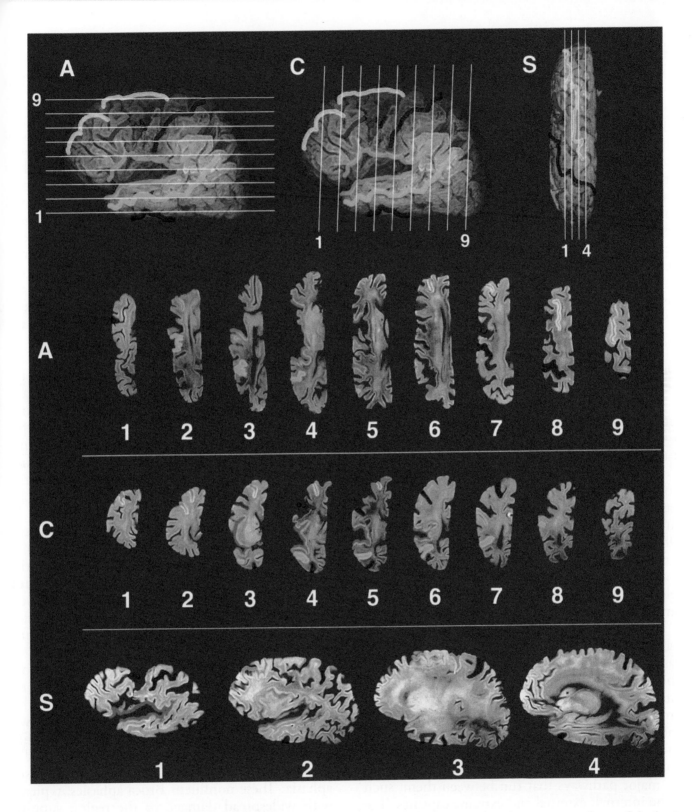

FIGURE 3 Magnetic resonance imaging (MRI) 3D reconstructions of the lateral and superior surfaces of the brain of Lelong with axial (A), coronal (C), and sagittal (S) sections through the brain. Note the wide sulci that suggest atrophy. Colors mark the central sulcus (dark blue), lateral fissure (aqua), inferior frontal sulcus (red), superior frontal sulcus (yellow), superior temporal sulcus (light green) and inferior temporal sulcus (brown). (Dronkers et al., Paul Broca's historic cases; high resolution MR imaging of the brains of Leborgne and Lelong, Brain, 2007, 130(5), pp. 1432-1441, by permission of Oxford University Press.)

Soon after, Broca had the opportunity to study another patient, the 84 year old Lelong, who could use only a few words when attempting to speak. This patient, also, had a lesion in the left ventrolateral frontal lobe in the region of the posterior part of the third frontal convolution (inferior frontal gyrus) (Fig. 1). Broca proceeded to report this case to the Anatomical Society of Paris and concluded that this region of the cerebrum was critical for speech (Broca, 1861c). He considered that comprehension of language was reasonably intact in these patients and that their main problem was speech production ("aphémie"). The term aphasia was coined later and, gradually, the region of the posterior part of the inferior frontal gyrus came to be referred to as Broca's area.

The brains of these two patients were not sectioned and Broca restricted his examination of the lesions to the surface of the brain. Although he considered the inferior frontal gyrus to be the critical region, he pointed out that, in the brain of Leborgne, the lesion included the short gyri of the insula and the corpus striatum as far as the lateral ventricle. The brains were preserved and, more recently, with the development of modern neuroimaging, the brain of Leborgne was scanned with computerized tomography (Castaigne et al., 1980; Signoret et al, 1984) and magnetic resonance imaging (Cabanis et al., 1994; Dronkers et al., 2007). The brain of Broca's second patient, Lelong, has also been scanned with magnetic resonance imaging by Dronkers and colleagues (2007).

The modern examination of the brain of Leborgne showed that, in the left hemisphere, there was significant damage to the inferior frontal gyrus, which extended deep into the brain to include the entire insula, the most anterior part of the superior temporal lobe, and rostral parts of the inferior parietal lobule (Fig. 2). The subcortical damage was extensive and included the claustrum, putamen, globus pallidus, head of the caudate nucleus and the major pathways that run between them, such as the internal, external and extreme capsules. The fibers that link the inferior parietal lobe to the frontal lobe, namely the superior longitudinal fasciculus were almost completely destroyed, as was the frontal-parietal periventricular white matter where axons from frontal and anterior parietal areas course to subcortical structures. In other words, there was massive disruption to the cortical and subcortical regions of the brain necessary for speech production.

Figure 3 shows sections from the magnetic resonance imaging (MRI) of the brain of Lelong, Broca's second patient (Dronkers et al., 2007). It is clear that the lesion is focused on the posterior part of the pars opercularis, with the inferior frontal gyrus anterior to this region spared. This is consistent with what can be observed from the surface of the brain (Fig. 1). The insula appears to be severely atrophied and there are small lesions in the region of the superior longitudinal fasciculus above the insula and lateral to the anterior horn of the lateral ventricle. Finally, there were abnormalities in the white matter pathways in the temporal lobe.

The recent examination of the brains of the two classic cases of Broca has shown that the lesions in these patients were extensive and clearly involved much more than the inferior frontal gyrus (Castaigne et al., 1980; Signoret et al, 1984; Cabanis et al., 1994; Dronkers et al., 2007). Thus, they do not provide evidence that damage limited to what came to be called Broca's area (posterior part of the inferior frontal gyrus) is sufficient for the clinical language impairment exhibited by these patients. Nevertheless, historically, these two cases focused attention on the ventrolateral frontal region of the left hemisphere of the brain as a critical zone for language production. Further definition of this critical zone that we now call Broca's area was not made until much later from electrical stimulation studies in patients undergoing neurosurgical operations.

Regardless of the definition of the critical zone for language processing in the posterior part of the inferior frontal gyrus (Broca's area), it is clear that a lesion restricted to this region of the brain will not cause the major aphasic problem that was first described by Broca and later came to be referred to as Broca's aphasia, nonfluent aphasia, or expressive aphasia. These nonfluent Broca aphasics, typically with widespread damage to the region supplied with blood by the upper division of the middle cerebral artery, exhibit effortful and dysprosodic speech, as well as severe agrammatism, i.e. severe reduction in the grammatical complexity of their sentences which are essentially reduced to content words (e.g., Goodglass, 1993; Grodzinsky, 2000).

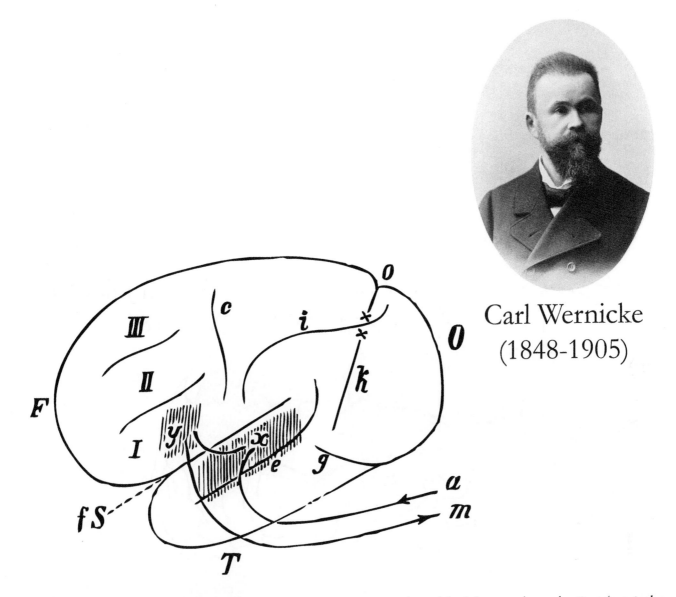

Carl Wernicke
(1848-1905)

FIGURE 4 Schematic diagram by Wernicke to illustrate his conceptual model of the speech mechanism (corticalen Sprachmechanism). The part of the superior temporal region of the cortex which he considers the auditory sensory speech center is linked across the Sylvian fissure with the motor speech region in the inferior frontal gyrus. Wernicke discussed the limited anatomical knowledge available at the time, including his own attempts at dissection, and concluded that the link between these two regions is probably made via the insula. Abbreviations: a, auditory pathway; c, central sulcus; e, parallel sulcus (superior temporal sulcus); F, frontal lobe; fS, fissure of Sylvius (lateral fissure); g, inferior occipital sulcus; i, intraparietal sulcus; k, anterior occipital sulcus; m, pathway to speech musculature; O, occipital lobe; o, parieto-occipital fissure; T, temporal lobe; x, sensory speech center; y, motor speech center; xy, association bundle between the two centers. I, inferior frontal gyrus; II, middle frontal gyrus; III, superior frontal gyrus. (From Wernicke, C., 1881, p. 205.)

When the lesion is more or less restricted to the posterior part of the inferior frontal gyrus without massive subcortical and insular damage, the patient will exhibit a much milder language production problem that resolves significantly with time (e.g., Penfield and Roberts, 1959; Mohr, 1976; Mohr et al., 1978). More recent studies taking advantage of modern neuroimaging methodology to define the lesions have also concluded that damage to the left frontal cortex involving Broca's area does not

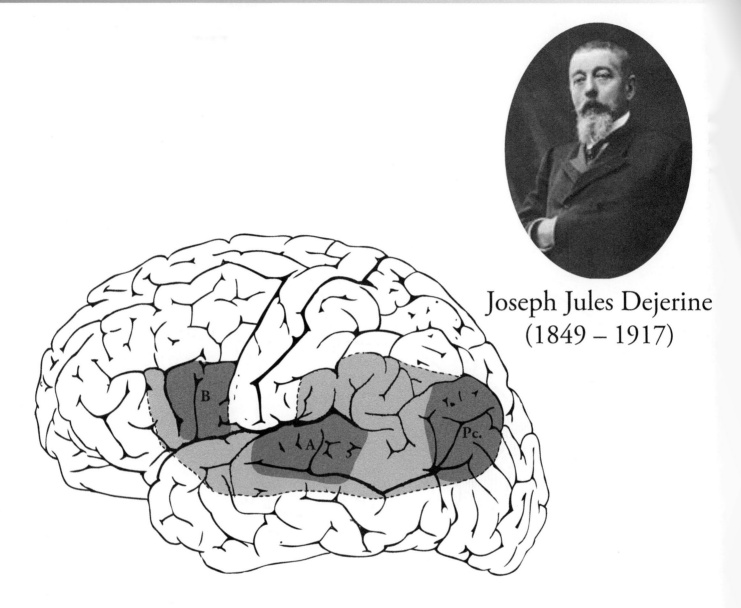

Joseph Jules Dejerine
(1849 – 1917)

FIGURE 5 The core language areas of the brain as depicted in Dejerine's famous textbook of neuroanatomy (1895). Abbreviations: A, focus of the auditory language region; B, Broca's region; Pc, angular gyrus.

necessarily result in nonfluent aphasia (e.g., Basso et al., 1985; Willmes and Poeck, 1993).

Dronkers and her colleagues have provided strong evidence that the anterior part of the insula in the left hemisphere is critical for the coordination of speech articulation (Dronkers, 1996; Borovsky et al., 2007; Baldo et al., 2011). This work has also received support from other studies (e.g., Nagao et al., 1999). Based on its close anatomical links to the precentral motor structures, it is reasonable to expect the anterior insula to be part of the orofacial motor system (e.g., Mesulam and Mufson,

1984). Functional neuroimaging studies have also provided evidence that articulation of speech involves the left anterior insula, in addition to the sensorimotor cortex (bilaterally) and subcortical sensorimotor structures, such as the basal ganglia and the cerebellum (e.g., Wise et al., 1999; Riecker et al., 2000; Nota and Honda, 2004). Since speech articulation involves the coordination of many muscles, one would expect the ventral precentral motor region, where the orofacial musculature is represented, and its intimately associated frontal opercular and anterior insular cortex to play a major role in this process.

The orbservations of Broca were followed by the seminal studies of Wernicke (1874, 1881) who posed the fundamental question of the critical region of the cerebral cortex leading to language comprehension impairment, in particular comprehension of spoken language. He argued that the critical region for auditory language comprehension was the superior temporal gyrus, including the cortex of the superior temporal sulcus and the adjacent lip of the middle temporal gyrus (Fig. 4). Wernicke was careful to point out that this superolateral temporal region was only the center of a much broader core language zone that included the whole peri-Sylvian cortical region and the insula. This was indeed a remarkable conception which proved to be consistent with the results of many studies of language impairments after lesions, as well as modern functional neuroimaging.

Following the influential observations and theoretical arguments of Wernicke, many other investigators attempted to define the critical region for language comprehension based on the study of aphasic impairments in patients (see, Bogen and Bogen, 1976, for an excellent discussion of the issue). Pierre Marie considered that the critical region for language, in the sense of a fundamental deficit to comprehend words and sentences after its damage, includes the supramarginal gyrus, the angular gyrus and the caudal part of the superior and middle temporal gyri (Marie, 1906). These early studies were largely based on older individuals with cerebrovascular accidents. Later, Marie and Foix (1917) studied soldiers sustaining brain injuries during the First World War and, again, concluded that the caudal part of the superior and middle temporal region, as well as the angular and supramarginal gyri, should be considered the critical region for language comprehension.

Recent research in which the lesions of aphasic patients were examined with modern structural neuroimaging has provided further evidence that the core language zone is a wide region of the cortex involving frontal, parietal and temporal areas surrounding the lateral fissure. For instance, semantic processing of both verbal and nonverbal auditory information, such as environmental sounds, was found to be impaired in aphasic patients with damage to the posterior middle and superior temporal gyri, as well as the inferior parietal lobule of the left hemisphere (Saygin et al., 2003). Moreover, recent lesion studies have confirmed the involvement of the middle temporal gyrus in the comprehension of words (e.g., Hart and Gordon, 1990; Hillis et al., 1999; Boatman et al., 2000; Bates et al., 2003; Dronkers et al., 2004; for a review, Binder, 2003) and functional neuroimaging with normal subjects has provided evidence for a role of the middle temporal region in word comprehension (e.g, Binder et al., 1997, 2009; Price, 2000, 2010).

THE POSTERIOR PARIETAL CORTEX AND LANGUAGE

Many of the classical lesion and electrical stimulation studies, including modern functional neuroimaging, have provided ample evidence for the importance of the supramarginal and angular gyri for various aspects of language processing. The seminal studies of Dejerine (1891a, 1891b, 1892) highlighted the importance of the posterior inferior parietal region to processes critical for reading and writing. Dejerine considered agraphia (writing impairment) to be a central problem of the expressive and receptive language disorders and disagreed with the attempt by Exner (1881) to introduce the idea of a special center for writing in the posterior middle frontal gyrus. Nevertheless, he considered that reading and writing could be selectively impaired by a lesion in the angular gyrus region of the language dominant hemisphere on the basis of two cases that he was able to examine in great detail (Dejerine, 1891a, 1891b, 1892) (Fig. 5). More recently, the contribution of the inferior parietal lobule to reading (Segal and Petrides, 2013) and syntactic processing (Bates et al., 1987; Stromswold et al., 1996; Caplan and Hildebrandt, 1998; Friederici et al., 2006; Friederici, 2009) has been emphasized. There is also excellent evidence for a role of the supramarginal gyrus in phonological processing (e.g., Saur et al., 2008; Church et al., 2011).

ELECTRICAL STIMULATION IN AWAKE PATIENTS UNDERGOING NEUROSURGICAL OPERATIONS

In addition to the attempts to correlate aphasic symptoms with lesion location, important information about the core language regions of the brain was provided by examination of the effects of electrical stimulation in alert, awake patients undergoing brain surgery under local anaesthesia. The aim of the electrical stimulation is to establish critical cortex for language, as indicated by interference with speech, in order to guide the limits of the surgical excision. The influential studies of Penfield and his colleagues in Montreal (Penfield and Rasmussen, 1950; Penfield and Roberts, 1959; Rasmussen and Milner, 1975) were continued by several other neurosurgeons (e.g., Ojemann and Whitaker, 1978; Ojemann, 1979, 1983, 1992; Ojemann et al., 1989; Duffau et al., 2005, 2008, 2009; Duffau, 2007, 2008; De Ribaupierre et al., 2012), leading to some of the best evidence regarding the critical cortical areas for language.

The patient may be asked to count or name pictures of objects or even be silent while electrical stimulation is applied to the region of interest. Stimulating the cortex of the precentral gyrus or the supplementary motor region in a silent patient may induce a vocal response in the form of a sustained or interrupted sound ("a sustained or interrupted vowel cry" according to Penfield and Roberts, 1959). This vocalization is produced from the lower part of the precentral gyrus where the orofacial region is represented and can be observed from stimulation both in the left and the right hemisphere (Fig. 6) (Penfield and Rasmussen, 1950). It is important to note that the vocal response is not any intelligible word, but rather a sound produced most probably as a result of the contraction of muscle groups following stimulation of motor areas controlling the orofacial musculature (Penfield and Boldrey, 1937; Woolsey et al., 1979; Fried et al., 1991; Martin et al., 2004).

Disruption of speech can be produced by electrical stimulation of the motor ventral precentral region, as well as the supplementary motor region, presumably due to interference with the motor production of words. In addition, interference with speech can be induced by stimulation in language specific regions of the cerebral cortex. Speech interference can take the form of cessation of speech (speech arrest) or interference as evidenced by hesitation, slurring of speech, word distortion, number confusion in a counting task, incorrect naming or failure to name objects presented, although speech per se may not be interrupted. Electrical stimulation provides evidence that language relevant processing has been interfered with and, thus, can help localize critical language cortex (Figs. 7 and 8).

It is interesting to note that speech interference and speech arrest can occur from stimulation anywhere in the peri-Sylvian cortex in the language dominant hemisphere. It includes the posterior part of the inferior frontal gyrus, the supramarginal and angular gyri of the inferior parietal lobule and, in the temporal lobe, the superior temporal gyrus, the superior temporal sulcus and the middle temporal gyrus. In other words, electrical stimulation in the core language regions of the cortex, as established by lesion studies, can lead to speech interference or speech arrest (Fig. 9).

The importance of the supplementary motor region for speech was also established by electrical stimulation studies. Penfield and Welch (1951) observed vocalization, interference and arrest of voluntary speech when stimulating the supplementary motor region in the language dominant

---

FIGURE 6 Sites from which vocalization could be evoked by electrical stimulation of the cortex by Penfield and colleagues. Vocalization could be evoked from stimulation of the ventral precentral region and the supplementary motor region on the medial surface of the hemisphere in both the right hemisphere (above) and the left hemisphere (below). Note that the points of stimulation evoking vocalization are primarily observed close to the central sulcus where they are more likely to involve the primary motor cortical region (area 4) and perhaps the closely related caudal area 6 (area 6VC). (Penfield, Wilder; Speech and Brain Mechanisms. 1959 Princeton University Press, 1987 renewed PUP Reprinted by Permission of Princeton University Press.) Color added to the original figure.

VOCALIZATION
Right Hemisphere

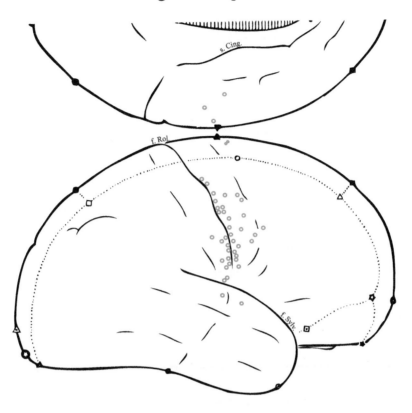

VOCALIZATION
Left Hemisphere

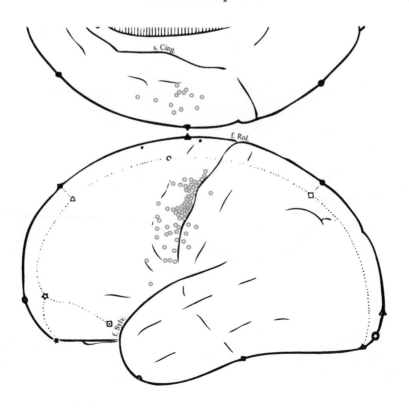

INTERFERENCE WITH SPEECH
(Left Hemisphere)

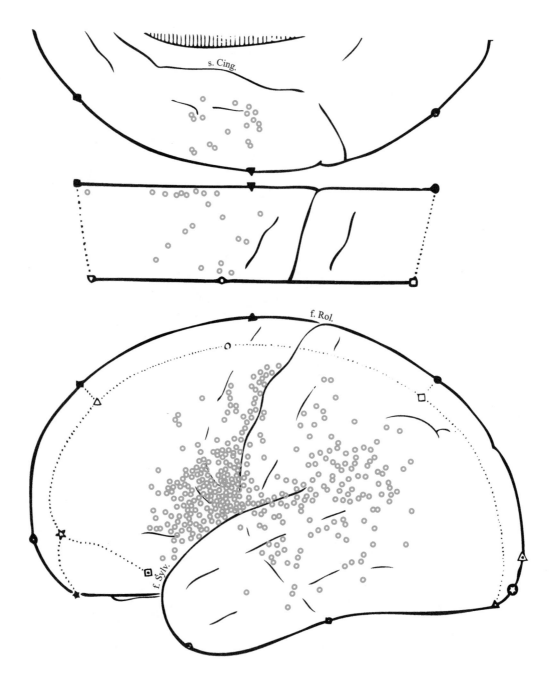

FIGURE 7 Sites in the left hemisphere leading to interference with speech following electrical stimulation. (Penfield, Wilder; Speech and Brain Mechanisms. 1959 Princeton University Press, 1987 renewed PUP Reprinted by Permission of Princeton University Press.) Color added to the original figure.

ARREST OF SPEECH
(Left Hemisphere)

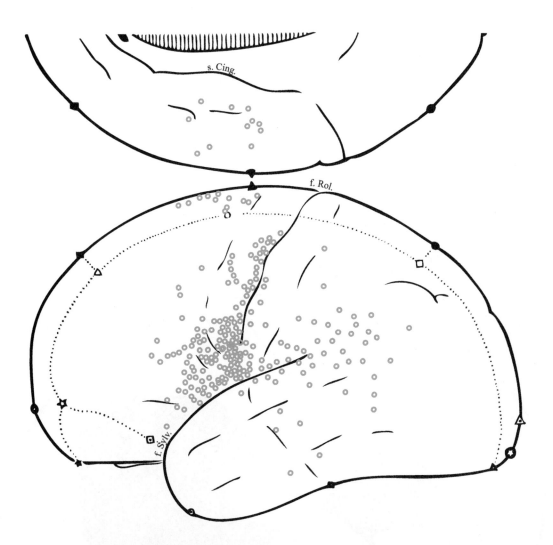

FIGURE 8 Sites in the left hemisphere leading to speech arrest following electrical stimulation. (Penfield, Wilder; Speech and Brain Mechanisms. 1959 Princeton University Press, 1987 renewed PUP Reprinted by Permission of Princeton University Press.) Color added to the original figure.

hemisphere and, on this basis, they suggested that this region must be involved in language. Several studies have subsequently shown that lesions involving this region of the cortex (although not always restricted to it) lead to reduction in language output (e.g., Goldberg, 1985; Rostomily et al., 1991; Krainik et al., 2003; Nachev et al., 2008; Chapados and Petrides, 2013). In addition

to the classical supplementary motor region, which can be sub-divided into supplementary proper and pre-supplementary motor areas, there are three somatotopically-organized motor areas just ventral and anterior to it that are likely to play some role in speech (Amiez and Petrides, 2012). Thus, it is possible that several motor areas found on the dorsomedial surface of the language domimant

SPEECH AREAS
EVIDENCE FROM STIMULATION
(Left Hemisphere)

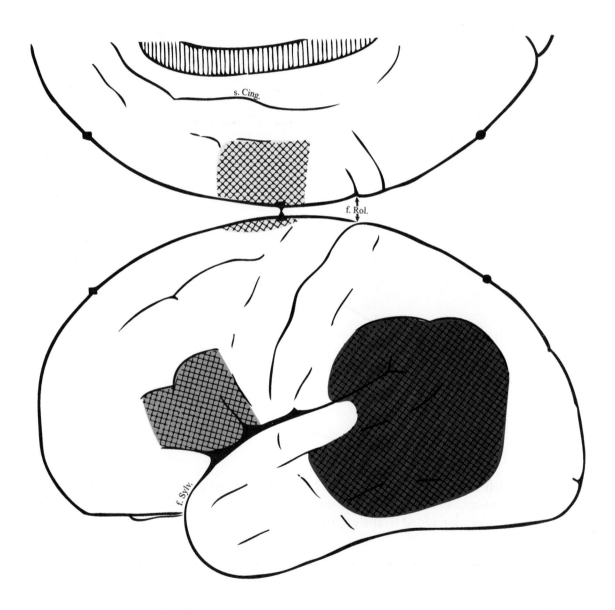

FIGURE 9 The anterior (green), posterior (purple), and medial (yellow) language regions of the left hemisphere based on evidence from electrical stimulation, according to Penfield and colleagues. (Penfield, Wilder; Speech and Brain Mechanisms. 1959 Princeton University Press, 1987 renewed PUP Reprinted by Permission of Princeton University Press.) Color added to the original figure.

hemisphere play a role in speech production (e.g., Paus et al., 1993).

Conclusion

The historical and modern studies examining aphasic symptoms in relation to lesion location and the electrical stimulation studies during neurosurgery have provided strong evidence that the core cortical language zone is the peri-Sylvian region. This region includes the cortical areas of the inferior frontal gyrus, the ventral parts of the precentral and postcentral cortex that control the orofacial musculature, the supramarginal and angular gyri of the inferior parietal lobe, and the superior, middle and to some extent the inferior temporal gyri. Broca's region, the ventral part of the precentral and postcentral cortex and the anterior supramarginal gyrus are in continuity with the cortex constituting the fronto-parietal operculum and these opercular areas are, in turn, in continuity with the cortex of the insula. Similarly, the superolateral temporal cortical areas (superior and middle temporal gyri) are in continuity with the temporal opercular region that includes Heschl's gyrus on which primary and secondary auditory cortex lies. More recently, a number of studies have addressed the possible role of the fusiform gyrus in certain aspects of reading (e.g., Cohen et al., 2000; Price and Devlin, 2003; Cohen and Dehaene, 2004). Finally, the dorsomedial region of the hemisphere that includes the supplementary motor region and the adjacent cingulate motor areas has been demonstrated to play a role in language production.

Morphological Features of the Core Language Regions: The Sulci and Gyri

Morphological Features of the Core Language Regions: The Sulci and Gyri

In this section of the atlas, an overview of the sulci and gyri of the peri-Sylvian region, where the core language areas are located, will be briefly presented. For a thorough discussion of all the sulci of the cerebral hemispheres, consult the atlas of the sulci and gyri of the human brain by Petrides (2012). For details of the sulcal patterns of the peri-Sylvian region and their variations, the reader should consult our published articles (e.g., Tomaiuolo et al., 1999; Germann et al., 2005; Zlatkina and Petrides, 2010; Petrides and Pandya, 2012; Segal and Petrides, 2012a).

The cerebral hemispheres of the human and non-human primates are traditionally divided into four lobes: the frontal, parietal, temporal, and occipital lobes. The central sulcus separates the frontal from the parietal lobe and the lateral fissure, also known as the Sylvian fissure, provides the natural boundary between the temporal and the frontal lobe (Fig. 10). It should be understood, however, that these separations are made for descriptive convenience and that the cortex is a continuous sheet of gray matter that is folded in order to fit into the space that is available inside the skull, creating the sulcal and gyral patterns that we observe. The schematic figures 10 and 11 (Petrides, 2012) display the typical pattern of the sulci and gyri of the lateral and medial surfaces of the human cerebral hemispheres. Variations in these patterns can be interpreted with careful examination.

As can be seen in figure 10, between the central sulcus (cs) and the inferior and superior precentral sulci lies the precentral gyrus (PrG), the motor region of the cerebral cortex. Immediately anterior to the ventral part of the precentral gyrus lies the inferior frontal gyrus (IFG), where Broca's region is located (Fig. 10). Posterior to the central sulcus, lies the postcentral gyrus (PoG) where the somatosensory representations are found (Fig. 10). The central sulcus continues medially and the medial extensions of the precentral and postcentral gyri form the morphological formation known as the paracentral lobule (PaCL; Fig. 11). The postcentral gyrus is succeeded posteriorly by the supramarginal gyrus (SmG; Fig. 10), which constitutes the rostral section of the inferior parietal lobule and, in turn, is succeeded caudally by the ill-defined morphological entity known as the angular gyrus (AnG; Fig. 10).

The cortex of the frontal and parietal lobes continues in the upper bank of the lateral fissure forming what is known as the fronto-parietal operculum (Fig. 12). The fronto-parietal opercular cortex is continuous with the insula, an island of cortex that is hidden deep in the lateral fissure. These relations can be appreciated in Figure 12 in which the lateral fissure in an MRI reconstruction has been opened to expose the fronto-parietal operculum and the insula.

Below the lateral fissure lies the superolateral temporal region, an essential component of the language cortex for auditory and multisensory comprehension of language. It comprises the superior temporal gyrus (STG), the cortex in the depth of the superior temporal sulcus (sts), and the middle temporal gyrus (MTG) that lies below the superior temporal sulcus (Fig. 10). Deep within the lower bank of the lateral fissure, approximately at the level of the postcentral gyrus, lies the transversely oriented Heschl's gyrus (highlighted in green), where the primary and secondary auditory cortical areas are located (Fig. 12).

INFERIOR FRONTAL GYRUS

The inferior frontal gyrus is delimited by the inferior frontal sulcus (ifs), dorsally, and the anterior part of the lateral (Sylvian) fissure, ventrally (Fig. 13). Three distinct parts of the inferior frontal gyrus can be identified: the pars opercularis (Op), the pars triangularis (Tr), and the pars orbitalis (Or). The caudal boundary of the pars opercularis is the inferior precentral sulcus (iprs) and its rostral

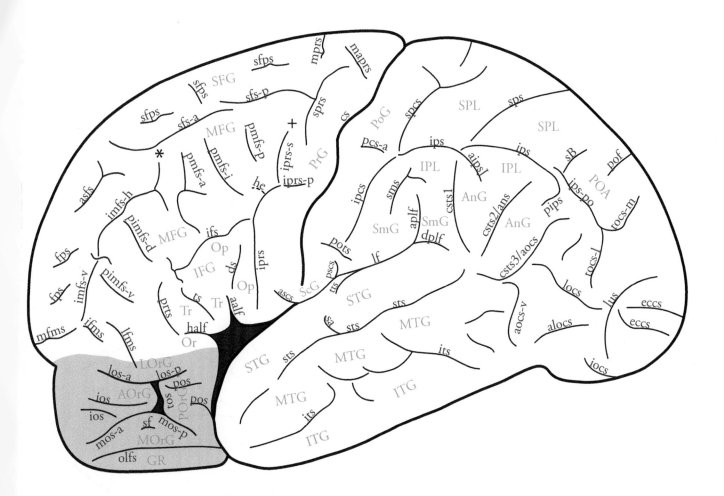

FIGURE 10 Schematic outline of the lateral surface of the hemisphere to illustrate the sulci and gyri of the human brain. The orbital surface of the frontal lobe is also displayed (grey shading). Abbreviations: aalf, ascending anterior ramus of the lateral fissure; aipsJ, anterior intermediate parietal sulcus of Jensen; alocs, accessory lateral occipital sulcus; AnG, angular gyrus; aocs-v, anterior occipital sulcus, ventral ramus; AOrG, anterior orbital gyrus; aplf, ascending posterior ramus of the lateral fissure; ascs, anterior subcentral sulcus; asfs, accessory superior frontal sulcus; cs, central sulcus; csts1, caudal superior temporal sulcus, first segment; csts2/ans, caudal superior temporal sulcus, second segment (angular sulcus); csts3/aocs, caudal superior temporal sulcus, third segment (anterior occipital sulcus); dplf, descending posterior ramus of the lateral fissure; ds, diagonal sulcus; eccs, external calcarine sulcus; fps, frontopolar sulcus; GR, gyrus rectus; half, horizontal ascending ramus of the lateral fissure; he, horizontal extension of the inferior precentral sulcus; IFG, inferior frontal gyrus; ifms, intermediate frontomarginal sulcus; ifs, inferior frontal sulcus; imfs-h, intermediate frontal sulcus, horizontal segment; imfs-v, intermediate frontal sulcus, vertical segment; iocs, inferior occipital sulcus; ios, intermediate orbital sulcus; ipcs, inferior post-central sulcus; ipcs-t, inferior post-central sulcus, transverse; IPL, inferior parietal lobule; iprs, inferior precentral sulcus; iprs-p, inferior precentral sulcus, posterior ramus; iprs-s, inferior precentral sulcus, superior ramus; ips, intraparietal sulcus; ips-po, intraparietal sulcus, paroccipital segment; ITG, inferior temporal gyrus; its, inferior temporal sulcus; lf, lateral fissure; lfms, lateral frontomarginal sulcus; locs, lateral occipital sulcus; LOrG, lateral orbital gyrus; los-a, lateral orbital sulcus, anterior ramus; los-p, lateral orbital sulcus, posterior ramus; lus, lunate sulcus; maprs, marginal precentral sulcus; MFG, middle frontal gyrus; mfms, medial frontomarginal sulcus; MOrG, medial orbital gyrus; mos-a, medial orbital sulcus, anterior ramus; mos-p, medial orbital sulcus, posterior ramus; mprs, medial precentral sulcus; MTG, middle temporal gyrus; olfs, olfactory sulcus; Op, opercular part of the inferior frontal gyrus; Or, orbital part of the inferior frontal gyrus; pcs-a, post-central sulcus, anterior; pimfs-d, paraintermediate frontal sulcus, dorsal; pimfs-v, paraintermediate frontal sulcus, ventral; pips, posterior intermediate parietal sulcus; pmfs-a, posterior middle frontal sulcus, anterior segment; pmfs-i, posterior middle frontal sulcus, intermediate segment; pmfs-p, posterior middle frontal sulcus, posterior segment; POA, parieto-occipital arcus; pof, parieto-occipital fissure; PoG, post-central gyrus; POrG, posterior orbital gyrus; pos, posterior orbital sulcus; pots, post-central transverse sulcus; PrG, precentral gyrus; prts, pretriangular sulcus; pscs, posterior subcentral sulcus; sa, sulcus acousticus; sB, sulcus of Brissaud; ScG, subcentral gyrus; sf, sulcus fragmentosus; SFG, superior frontal gyrus; sfps, superior frontal paramidline sulcus; sfs-a, superior frontal sulcus, anterior segment; sfs-p, superior frontal sulcus, posterior segment; SmG, supramarginal gyrus; sms, supramarginal sulcus; spcs, superior post-central sulcus; SPL, superior parietal lobule; sprs, superior precentral sulcus; sps, superior parietal sulcus; STG, superior temporal gyrus; sts, superior temporal sulcus; tocs-l, transverse occipital sulcus, lateral ramus; tocs-m, transverse occipital sulcus, medial ramus; tos, transverse orbital sulcus; Tr, triangular part of the inferior frontal gyrus; ts, triangular sulcus; tts, transverse temporal sulcus.

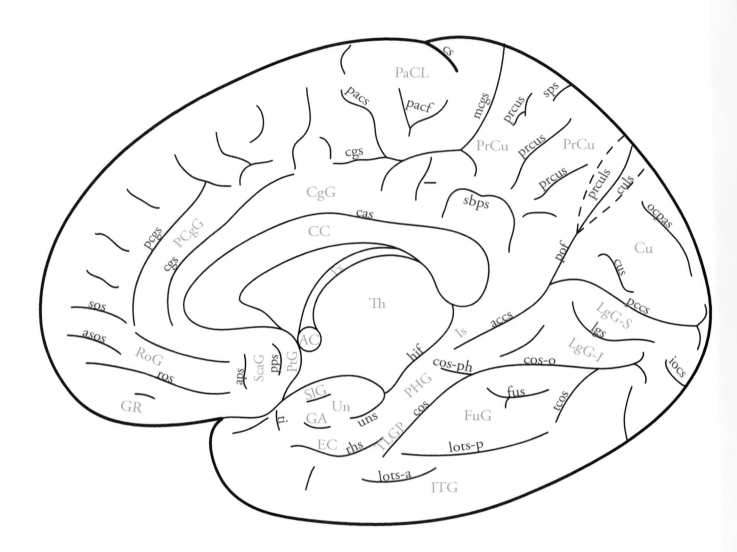

FIGURE 11 Schematic outline of the medial surface of the hemisphere to illustrate the sulci and gyri of the human brain. Abbreviations: AC, anterior commissure; accs, anterior calcarine sulcus; aps, anterior parolfactory sulcus; asos, accessory supra-orbital sulcus; cas, callosal sulcus; CC, corpus callosum; CgG, cingulate gyrus; cgs, cingulate sulcus; cos, collateral sulcus; cos-o, collateral sulcus, occipital ramus; cos-ph, collateral sulcus, parahippocampal extension; cs, central sulcus; Cu, cuneus; culs, cuneal limiting sulcus; cus, cuneal sulcus; EC, entorhinal cortex; FuG, fusiform gyrus; fus, fusiform sulcus; Fx, fornix; GA, gyrus ambiens; GR, gyrus rectus; hif, hippocampal fissure; Is, isthmus; ITG, inferior temporal gyrus; LgG-I, inferior lingual gyrus; LgG-S, superior lingual gyrus; lgs, lingual sulcus; lots, lateral occipitotemporal sulcus; mcgs, marginal ramus of the cingulate sulcus; ocpas, occipital paramedial sulcus; pacf, paracentral fossa; PaCL, paracentral lobule; pacs, paracentral sulcus; pccs, posterior calcarine sulcus; PCgG, paracingulate gyrus; pcgs, paracingulate sulcus; PHG, parahippocampal gyrus; pof, parieto-occipital fissure; pps, posterior parolfactory sulcus; PrCu, precuneus; prculs, precuneal limiting sulcus; prcus, precuneal sulcus; PtG, paraterminal gyrus; rhs, rhinal sulcus; RoG, rostral gyrus; ros, rostral sulcus; sbps, subparietal sulcus; ScaG, subcallosal gyrus; SlG, semilunar gyrus; sos, supra-orbital sulcus; sps, superior parietal sulcus; tcos, transverse collateral sulcus; Th, thalamus; ti, temporal incisure; TLGP, temporo-limbic gyral passage; Un, uncus; uns, uncal sulcus.

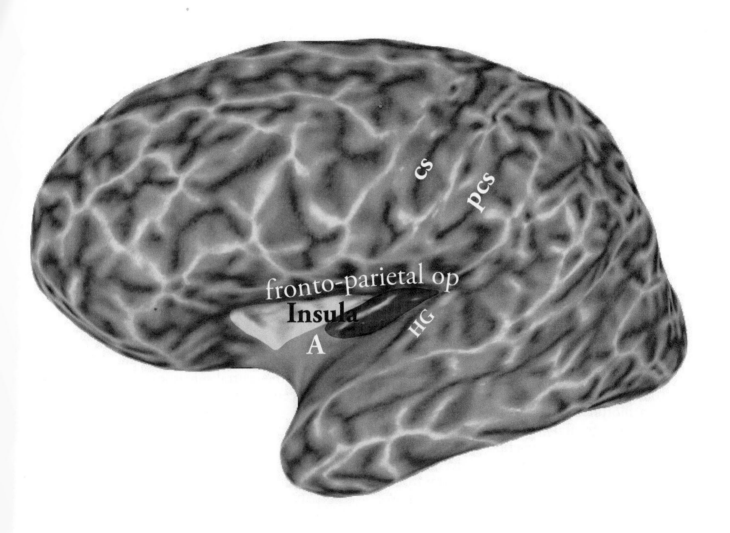

FIGURE 12 Three dimensional reconstruction of the magnetic resonance image (MRI) of a human brain in which the sulci have been opened so that the fronto-parietal operculum and the insula can be viewed. The three short gyri of the insula (cytoarchitectonically dysgranular insula) are marked in yellow and the two long gyri of the insula (cytoarchitectonically granular insula) are marked in red. The apex (pole) of the insula is marked with the letter A. The apex is the cytoarchitectonically agranular part of the insula. Heschl's gyrus is marked in green. Abbreviations: A, apex of the insula; cs, central sulcus; pcs, postcentral sulcus; HG, Heschl's gyrus.

---

boundary is the ascending anterior ramus of the lateral fissure (aalf) (Fig. 13). In some brains, an additional sulcus, the diagonal sulcus (ds) may be identified within the pars opercularis (Fig. 13). The diagonal sulcus, when present, may remain separate or may blend with the ascending sulcus.

The pars triangularis (Tr) lies anterior to the pars opercularis (Op) and is delimited by the ascending (aalf) and the horizontal (half) rami of the lateral fissure. The triangular sulcus (ts), also known as the incisura capitis, separates the pars triangularis into an anterior and a posterior part (Fig. 13). The inferior frontal sulcus, which originates, posteriorly, close to the inferior precentral sulcus, ends approximately at the anterior half of the pars triangularis. Its anterior terminal branch may merge superficially with the triangular sulcus. The inferior frontal sulcus may sometimes separate into a posterior and an anterior part. It is sometimes stated, incorrectly, that the inferior frontal sulcus may continue, ventrally, almost as far as the lateral margin of the orbital frontal region. This false impression is the result of the blending of the anterior part of the inferior frontal sulcus with a distinct sulcus that often forms the anterior end of the pars triangularis, the pretriangular sulcus (prts) (Petrides, 2012). Below the horizontal sulcus, on the ventralmost part of the lateral surface of the hemisphere and extending onto the orbital suface as far as the caudal part of the lateral orbital sulcus (los-p), lies the pars orbitalis (Or) of the inferior frontal gyrus (Fig. 13).

Anterior to the termination of the inferior frontal sulcus (i.e. above and also anterior to the pars triangularis of the inferior frontal gyrus), extends a morphologically ill-defined region, the anterior middle frontal gyrus. Although not a core language region, a brief description of its sulci is provided here because some of the sulci can blend with those of the anterior part of the inferior frontal gyrus. It should be noted that such anastomoses are superficial and the end of one sulcus and the beginning of the other can clearly be established by examining the depths of the sulci. The intermediate frontal sulcus courses, initially, in an approximately horizontal direction and, then, turns ventrally to reach the medial frontomarginal sulcus in the frontal pole, thus forming a more or less horizontal ramus (imfs-h) and a vertical ramus (imfs-v) (Fig. 10). We have investigated the morphology of the anterior middle frontal region intensively. Two main sulci can be identified lateral to the intermediate frontal sulcus and we named these sulci the dorsal and ventral paraintermediate frontal sulci (pimfs-d, pimfs-v) because they lie immediately adjacent to the intermediate frontal sulcus (Petrides and Pandya, 2012). The dorsal paraintermediate frontal sulcus (pimfs-d) originates close to the horizontal segment of the intermediate frontal sulcus (imfs-h) and, ventrally, is directed towards the dorsal part of the pars triangularis (Fig. 10). It may merge, superficially, with sulci that terminate in this region, such as the triangular sulcus or the rostral end of the inferior frontal sulcus. The ventral paraintermediate frontal sulcus (pimfs-v) originates near the ventral segment of the intermediate frontal sulcus and may blend, superficially, with the pretriangular sulcus (prts) and the lateral frontomarginal sulcus (lfms).

VENTRAL PRECENTRAL REGION

The central sulcus (the sulcus of Rolando) forms the boundary between the frontal and the parietal lobes on the lateral and medial surfaces of the cerebral hemispheres (Figs. 10 and 11). Near the lateral fissure, the central sulcus is bounded by two short sulci, the anterior subcentral sulcus (ascs) and the posterior subcentral sulcus (pscs), forming the subcentral gyrus (ScG) (Fig. 10). The subcentral gyrus may lie within the lateral fissure, giving the impression that the central sulcus joins the lateral fissure. The precentral motor region of the brain extends from the anterior bank of the central sulcus to the superior and inferior precentral sulci, which form the anterior limit of the precentral gyrus (Fig. 10). The inferior precentral sulcus (iprs) separates the ventral orofacial part of the precentral gyrus from the pars opercularis of the inferior frontal gyrus. It continues dorsally for a considerable distance so that its superior ramus (iprs-s) forms the posterior margin of the middle frontal gyrus. A horizontally directed protrusion of the inferior precentral sulcus, the horizontal extension (he), projects into the middle frontal gyrus, but this short sulcus is often submerged and cannot be observed on the surface of the brain (dotted line in Fig. 10). A short sulcus close to the inferior precentral sulcus frequently projects towards the central sulcus, the posterior ramus of the inferior precentral sulcus (iprs-p) (Fig. 10).

DORSOMEDIAL MOTOR REGION: PARACENTRAL LOBULE AND THE SUPPLEMENTARY MOTOR REGION

The central sulcus (cs) reaches the superior margin of the hemisphere and, often, continues medially to end just anterior to the marginal ramus of the cingulate sulcus (mcgs) (Fig. 11). The medial

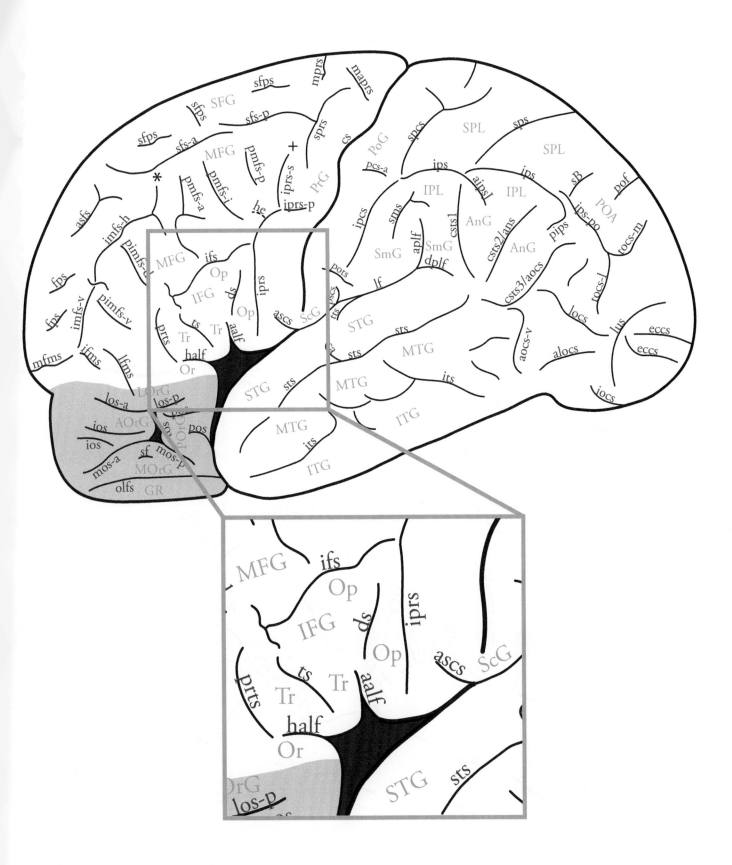

FIGURE 13 Schematic diagram of the lateral surface of the left hemisphere in which the region of the inferior frontal gyrus has been expanded to illustrate the sulci. For abbreviations, see Abbreviations List.

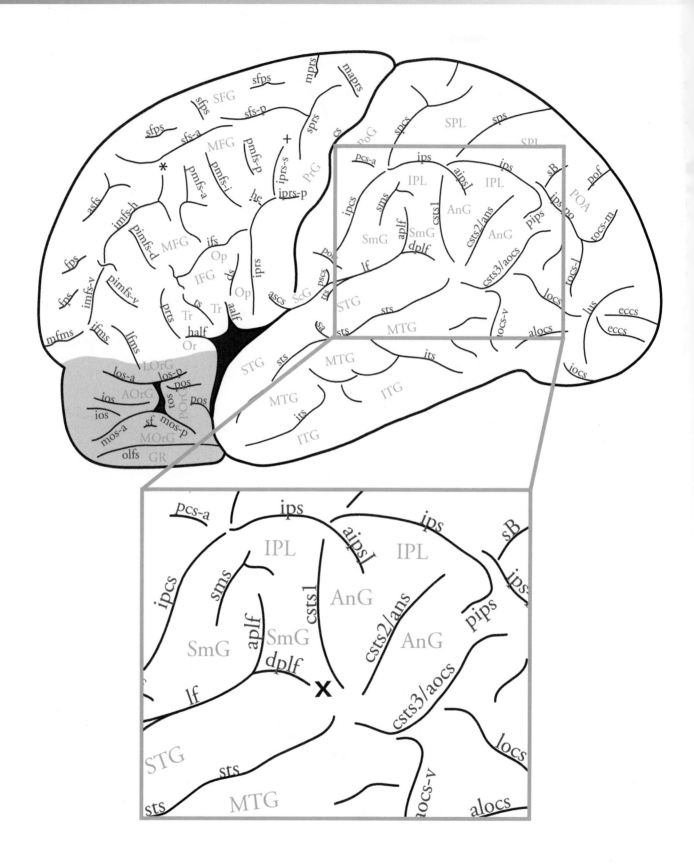

FIGURE 14 Schematic diagram of the lateral surface of the left hemisphere with region of the inferior parietal lobule expanded to illustrate the sulci. x refers to the parieto-temporal isthmus which is the narrow passage between the posterior supramarginal gyrus and the posterior temporal gyrus. For abbreviations, see Abbreviations List.

extension of the precentral motor region and the postcentral somatosensory region form the paracentral lobule (PaCL) which is limited, posteriorly, by the marginal ramus of the cingulate sulcus (Fig. 11). The shallow dimple within the paracentral lobule, the paracentral fossa (pacf), must not be confused with the paracentral sulcus (pacs) which forms the anterior margin of the paracentral lobule. The paracentral lobule consists of the medial extensions of the precentral and postcentral gyri where the leg region is represented. Anterior to the paracentral lobule lies the supplementary motor region, where another representation of the body was found by Penfield and colleagues. Electrical stimulation of this region gives rise to vocalization and speech arrest (see Figs. 6-9).

LATERAL PARIETAL REGION

The anterior part of the parietal lobe, namely the postcentral gyrus (PoG), is the purely somatosensory region of the parietal lobe where somatic representations of the various body parts are found. Its posterior boundary is formed by the superior and inferior postcentral sulci (spcs, ipcs) (Fig. 14). The posterior parietal region that extends behind the postcentral gyrus is divided into a superior and an inferior parietal lobule (SPL, IPL) by the intraparietal sulcus (ips) (Fig. 14). Two major morphological entities constitute the inferior parietal lobule: the supramarginal gyrus (SmG) and the angular gyrus (AnG). The supramarginal gyrus is an inverted U shaped convolution that is formed around the ascending posterior ramus of the lateral fissure (aplf) (Fig. 14). A variable sulcus, the supramarginal sulcus (sms), is often encountered within the supramarginal gyrus.

The angular gyrus is a poorly defined region that spreads around the caudal rami of the superior temporal sulcus. These rami originate near the temporal part of the superior temporal sulcus and ascend into the posterior part of the inferior parietal lobule (see Petrides, 2012; Segal and Petrides, 2012a). The first caudal superior temporal sulcus (csts1) is often confused with the anterior intermediate parietal sulcus of Jensen (aipsJ). The reason for this confusion is the fact that they approach each other near the intraparietal sulcus (Fig. 14).

These two sulci, however, can easily be distinguished: the anterior intermediate sulcus of Jensen emerges from within the intraparietal sulcus and terminates, usually, posterior to the superior end of the first ascending segment of the caudal superior temporal sulcus (Fig. 14). If these sulci blend, the confluence is superficial.

The inferior parietal lobule (IPL) continues posteriorly into the occipital region and ventrally into the posterior temporal region. Because the posterior parietal region and the posterior temporal region merge imperceptibly and, in the language dominant hemisphere form the posterior language zone, a number of interesting points must be highlighted here. First, the posterior end of the supramarginal gyrus (SmG) and the superior temporal gyrus are continuous. The x symbol in the inset of figure 14 is intended to highlight this lateral parieto-temporal isthmus (ισθμός, a narrow passage in Greek). Cytoarchitectonically, it is the point where the parietal cortex of the posterior supramarginal gyrus merges with the posterior temporal cortex, both of which are related cytoarchitectonic areas (see Cytoarchitecture section of the atlas). Similarly, the cortex in the angular gyrus region is in continuity with the cortex of the superior temporal sulcus and the middle temporal gyrus.

LATERAL TEMPORAL REGION

The primary auditory cortex and the surrounding secondary cortical region lie on a distinct morphological entity, Heschl's gyrus (HG). This is a transverse gyrus that is found on the ventral bank of the lateral fissure. In figure 15, one can observe Heschl's gyrus and the surrounding sulci and gyri in detail. This figure provides a photograph of the superolateral temporal lobe in which the frontoparietal operculum has been removed to expose Heschl's gyrus. The sulcus of Heschl (sH) separates Heschl's gyrus from the posterior cortical region on the hidden surface of the temporal lobe. A sulcus slightly posterior to the sulcus of Heschl, the transverse temporal sulcus (tts), may reach the surface of the superior temporal gyrus (STG). Another short sulcus, the sulcus acousticus (sa), may also appear cutting into the superior temporal gyrus as

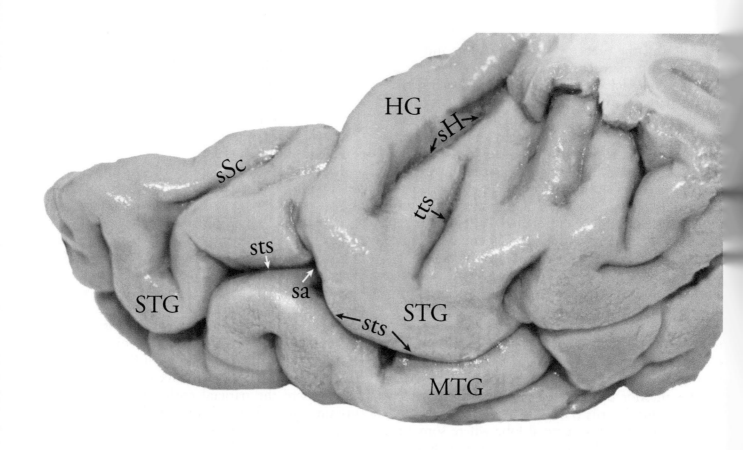

FIGURE 15 Photograph of a cadaver human brain in which the fronto-parietal operculum has been removed to illustrate Heschl's gyrus, the planum polare and planum temporale that are hidden within the lateral fissure. Abbreviations: HG, Heschl's gyrus; MTG, middle temporal gyrus; sH, sulcus of Heschl; STG, superior temporal gyrus; sa, sulcus acousticus; sSc, sulcus of Schwalbe; sts, superior temporal sulcus; tts, transverse temporal sulcus.

---

an extension of the superior temporal sulcus (sts) in front of the transverse temporal sulcus. When present on the lateral surface of the superior temporal gyrus, these two sulci provide an indication of the location of Heschl's gyrus within the lateral fissure. The cortex anterior to Heschl's gyrus on the superior surface of the temporal lobe is referred to as the temporal planum polare, within which one or more sulci of Schwalbe (sSc) can be identified. The sulcus separating Heschl's gyrus from the cortex that lies anterior to it is sometimes referred to as the first transverse temporal sulcus (Rademacher et al., 1993). The cortex posterior to Heschl's gyrus

on the superior surface of the temporal lobe may be folded into an additional but less prominent convolution, which has led some investigators to refer to a second Heschl's gyrus.

The superior temporal sulcus is very long and proceeds in an inclined rostro-dorsal direction, starting near the temporal pole and terminating in the inferior parietal lobule (Fig. 10). Like most of the long sulci of the hemisphere, it is not a simple fold of the cortex, but rather a complex of sulcal elements that define the borders of many different parts of the brain. The most anterior part of the superior temporal sulcus is independent of its main part in the temporal lobe, which runs ventral to the lateral fissure, in a more or less parallel direction, leading many of the classical investigators to refer to the superior temporal sulcus as the parallel sulcus (Shellshear, 1927). The superior temporal sulcus forms the ventral border of the superior temporal gyrus and the dorsal border of the middle temporal gyrus (Fig. 10). The three caudal branches of the superior temporal sulcus are discussed together with the sulci of the inferior parietal lobule because they course primarily in this region of the brain and define its morphology.

Below the superior temporal sulcus, there is a series of sulci that are nowadays named collectively as the inferior temporal sulcus (Fig. 10). In the older anatomical literature, the sulcal branches closest to the superior temporal sulcus were often referred to as the middle temporal sulcus, but this term has virtually disappeared from the recent literature.

INFEROMEDIAL TEMPORAL REGION

The inferomedial region of the temporal lobe is morphologically marked by the parahippocampal gyrus (PHG) (Fig. 11) which comprises phylogenetically older periallocortical and proisocortical areas. The border between these phylogenetically older cortical areas and the isocortical areas encountered on the adjacent inferomedial temporal region is formed by three sulci: the temporal incisure (ti), the rhinal sulcus (rhs), and the collateral sulcus (cos). The collateral sulcus, which marks the lateral boundary of the posterior parahippocampal region, was referred to as the occipitotemporal sulcus in the older anatomical literature (e.g., Economo and Koskinas, 1925). It is a long sulcus that continues, posteriorly, all the way to the occipital region where its occipital component (cos-o) forms the ventral border of the lingual gyrus (Fig. 11).

Lateral to the parahippocampal gyrus and anterior to the lingual gyrus lies the fusiform gyrus (FuG), also known as the occipito-temporal gyrus. The medial border of this gyrus is formed by the collateral sulcus (also known as the medial occipitotemporal sulcus) and its lateral border by the lateral occipitotemporal sulcus (lots) (Fig. 11). The lateral occipitotemporal sulcus is sometimes referred to as the "occipitotemporal sulcus" but this term can be ambiguous because it has also been used, in the classical anatomical literature, to refer to the collateral sulcus (i.e. the medial occipitotemporal sulcus).

THE INSULA

The insula is an island of cortex hidden within the lateral fissure and is in continuity with the fronto-parietal opercular region (e.g., Mesulam and Mufson, 1984; Türe et al., 1999). It is a region of the cortex that appears to play a major role in motor aspects of speech (Ackermann and Riecker, 2004). The insula (the island of Reil) comprises five convolutions that are disposed in the form of a fan. The ventrally located apex (A) of the fan is known as the pole of the insula (Fig. 12). Ventrally, the insula merges with the temporal lobe and, rostrally, with the orbital region of the frontal lobe, the so-called fronto-orbital operculum. It is separated from the surrounding opercular structures by the circular insular sulcus (the limiting insular sulcus). The insula comprises three short gyri, anteriorly, and two long gyri, posteriorly: gyrus brevis I, gyrus brevis II, gyrus brevis III, gyrus longus I, and gyrus longus II (Fig. 16). The central insular sulcus (cis) courses approximately at the anteroposterior level of the central sulcus. It separates the anterior three short gyri from the two long gyri, in other words the anterior insula from the posterior insula. The post-central insular sulcus (pcis) separates gyrus longus I from gyrus longus II. Gyrus brevis II is separated from gyrus brevis III by the precentral insular sulcus (pris) and gyrus brevis I is separated from gyrus brevis II by sulcus brevis II (sbi II) (Fig. 16). Sulcus brevis I separates gyrus brevis I from the accessory short insular gyrus, which lies between the rostral end of the insula and the fronto-orbital operculum.

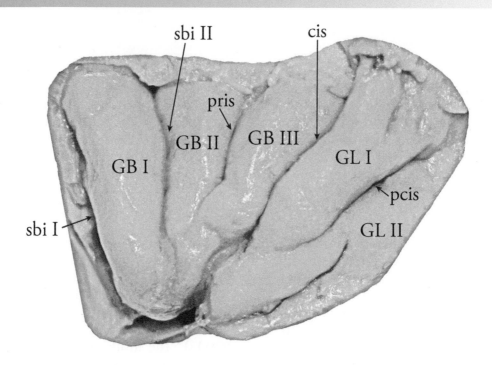

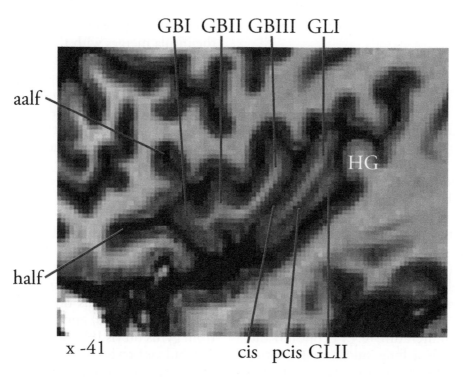

FIGURE 16 Photograph of the exposed insula of the left hemisphere. The fronto-parietal operculum, the orbitofrontal operculum, and the temporal operculum have been removed to expose the insula. The gyri of the insula can be seen in appropriate sections of magnetic resonance images of the brain. At the appropriate medio-lateral level, a sagittal section allows a view of the gyri of the insula (sagittal section, x -41). Similarly, the gyri of the insula can be viewed at an appropriate dorsoventral level (horizontal section, z -3). At the appropriate antero-posterior levels, coronal sections (y +17 to y -19) allow the investigator to examine, individually, the various gyri of the insula. Abbreviations: cis, central insular sulcus; GB I, gyrus brevis insulae (anterior); GB II, gyrus brevis insulae (middle); GB III, gyrus brevis insulae (posterior); GL I, gyrus longus insulae (anterior); GL II, gyrus longus insulae (posterior); pcis, post-central insular sulcus; pris, precentral insular sulcus; sbi I, sulcus brevis insulae (first); sbi II, sulcus brevis insulae (second).

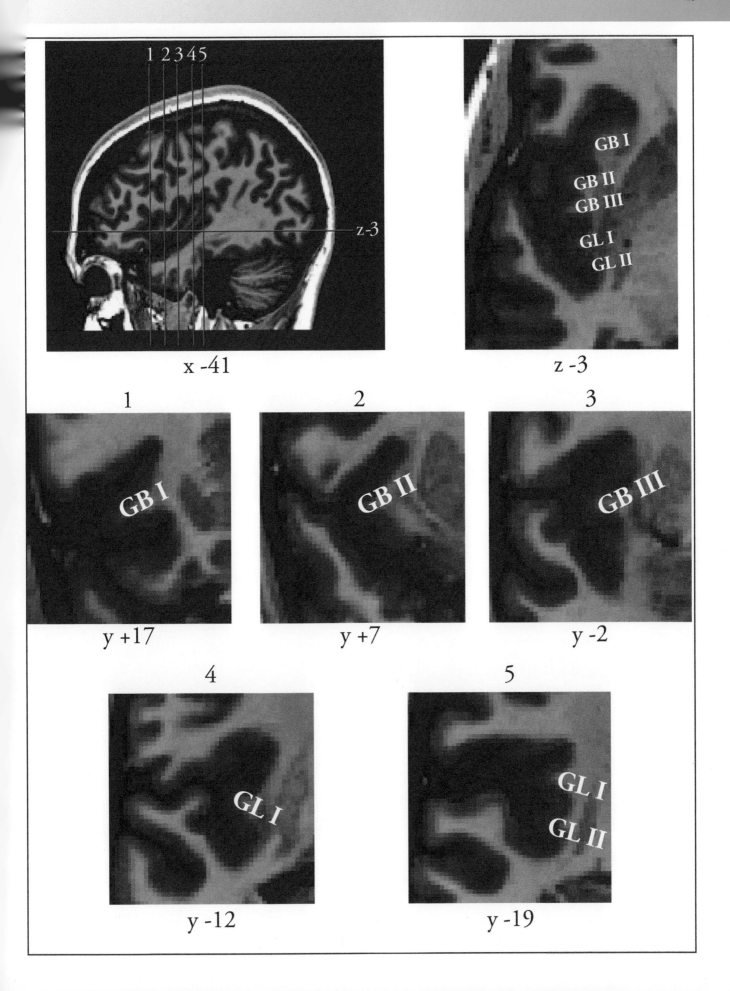

x -41

z -3

1

GB I

y +17

2

GB II

y +7

3

GB III

y -2

4

GL I

y -12

5

GL I
GL II

y -19

Abbreviations

A

A	amygdala
aalf	ascending anterior ramus of the lateral fissure
AC	anterior commissure
accs	anterior calcarine sulcus
aipsJ	anterior intermediate parietal sulcus of Jensen
alocs	accessory lateral occipital sulcus
AnG	angular gyrus
ans	angular sulcus
aocs	anterior occipital sulcus
aocs-v	anterior occipital sulcus, ventral ramus
AOrG	anterior orbital gyrus
aplf	ascending posterior ramus of the lateral fissure
aps	anterior parolfactory sulcus
ascs	anterior subcentral sulcus
asfs	accessory superior frontal sulcus
ASIG	accessory short insular gyrus
asos	accessory supra-orbital sulcus

C

cas	callosal sulcus
CC	corpus callosum
CC-B	corpus callosum, body
CC-G	corpus callosum, genu
CC-R	corpus callosum, rostrum
CC-Sp	corpus callosum, splenium
Cd	caudate nucleus
CgG	cingulate gyrus
cgs	cingulate sulcus
Cl	claustrum
cis	central insular sulcus
cos	collateral sulcus (medial occipitotemporal sulcus)
cos-o	collateral sulcus, occipital ramus (occipital collateral sulcus)
cos-ph	collateral sulcus, parahippocampal extension
cris	circular insular sulcus
cs	central sulcus (sulcus of Rolando)
csts1	caudal superior temporal sulcus, first segment
csts2/ans	caudal superior temporal sulcus, second segment (angular sulcus)

csts3/aocs	caudal superior temporal sulcus, third segment (anterior occipital sulcus)
Cu	cuneus
CuG	cuneal gyrus
culs	cuneal limiting sulcus (sulcus paracalcarinus of Elliot Smith)
cus	cuneal sulcus (sulcus sagittalis inferior cunei of Retzius)

D

dplf	descending posterior ramus of the lateral fissure
ds	diagonal sulcus

E

EC	entorhinal cortex
eccs	external calcarine sulcus (sulcus calcarinus externus of Cunningham)

F

fps	frontopolar sulcus
ftts	first transverse temporal sulcus
FuG	fusiform gyrus
fus	fusiform sulcus
Fx	fornix

G

GA	gyrus ambiens
GB	gyrus brevis insulae (short insular gyrus)
GB I	gyrus brevis insulae, anterior or first (short insular gyrus, anterior or first)
GB II	gyrus brevis insulae, middle or second (short insular gyrus, middle or second)
GB III	gyrus brevis insulae, posterior or third (short insular gyrus, posterior or third)
GL	gyrus longus insulae (long insular gyrus)
GL I	gyrus longus insulae, anterior or first (long insular gyrus, anterior or first)
GL II	gyrus longus insulae, posterior or second (long insular gyrus, posterior or second)
GP	globus pallidus
GR	gyrus rectus

H

half   horizontal ascending ramus of the lateral fissure

he   horizontal extension of the inferior precentral sulcus

HG   Heschl's gyrus

Hi   hippocampus

hif   hippocampal fissure

I

ICuA   intercuneate arcus (arcus intercuneatus of Elliot Smith)

IFG   inferior frontal gyrus

ifms   intermediate frontomarginal sulcus

ifs   inferior frontal sulcus

ifs-t   inferior frontal sulcus, terminal ramus

imfs-h   intermediate frontal sulcus, horizontal segment (middle frontal sulcus)

imfs-v   intermediate frontal sulcus, vertical segment (middle frontal sulcus)

iocs   inferior occipital sulcus

ios   intermediate orbital sulcus

ipcs   inferior post-central sulcus

ipcs-t   inferior post-central sulcus, transverse

IPL   inferior parietal lobule

iprs   inferior precentral sulcus

iprs-p   inferior precentral sulcus, posterior ramus

iprs-s   inferior precentral sulcus, superior ramus

ips   intraparietal sulcus

ips-po   intraparietal sulcus, paroccipital segment (paroccipital sulcus)

iros   inferior rostral sulcus

Is   isthmus

ITG   inferior temporal gyrus

its   inferior temporal sulcus

L

lf   lateral fissure

lfms   lateral frontomarginal sulcus

LgG   lingual gyrus

LgG-I   inferior lingual gyrus

LgG-S   superior lingual gyrus

lgs   lingual sulcus (intralingual sulcus)

locs   lateral occipital sulcus (prelunate sulcus)

LOrG   lateral orbital gyrus

los   lateral orbital sulcus

los-a   lateral orbital sulcus, anterior ramus

los-p   lateral orbital sulcus, posterior ramus

lots   lateral occipitotemporal sulcus

lus   lunate sulcus

M

maprs   marginal precentral sulcus

mcgs   marginal ramus of the cingulate sulcus

MFG   middle frontal gyrus

mfms   medial frontomarginal sulcus

MOrG   medial orbital gyrus

mos   medial orbital sulcus

mos-a   medial orbital sulcus, anterior ramus

mos-p   medial orbital sulcus, posterior ramus

mprs   medial precentral sulcus

MTG   middle temporal gyrus

O

ocpas   occipital paramedial sulcus (paramesial sulcus, superior sagittal sulcus of the cuneus)

olfs   olfactory sulcus

Op   opercular part of the inferior frontal gyrus (pars opercularis)

Or   orbital part of the inferior frontal gyrus (pars orbitalis)

P

pacf   paracentral fossa

PaCL   paracentral lobule

pacs   paracentral sulcus

pccs   posterior calcarine sulcus

PCgG   paracingulate gyrus

pcgs   paracingulate sulcus

pcis   post-central insular sulcus

pcs   post-central sulcus

pcs-a   post-central sulcus, anterior

PHG   parahippocampal gyrus

pimfs   paraintermediate frontal sulcus

pimfs-d   paraintermediate frontal sulcus, dorsal

pimfs-v   paraintermediate frontal sulcus, ventral

pips   posterior intermediate parietal sulcus

pmfs-a   posterior middle frontal sulcus, anterior segment

pmfs-i   posterior middle frontal sulcus, intermediate segment

pmfs-p   posterior middle frontal sulcus, posterior segment

POA   parieto-occipital arcus (arcus parieto-occipitalis, parieto-occipital arch)

PoG   post-central gyrus

poi   parieto-occipital incisure (incisura parieto-occipitalis of Elliot Smith)

POrG   posterior orbital gyrus

pof   parieto-occipital fissure

pos   posterior orbital sulcus

pots   post-central transverse sulcus

ppacs   pre-paracentral sulcus

ppcs   posterior ramus of the post-central sulcus

pps   posterior parolfactory sulcus

prcs   precentral sulcus

PrCu   precuneus

prculs   precuneal limiting sulcus

prcus   precuneal sulcus

PrG   precentral gyrus

pris   precentral insular sulcus

prts   pretriangular sulcus (sulcus radiatus of Eberstaller)

pscs   posterior subcentral sulcus

PT   planum temporale

PtG   paraterminal gyrus

Pu   putamen

R

rhs   rhinal sulcus (anterior collateral sulcus)

RoG   rostral gyrus

ros   rostral sulcus

S

sa   sulcus acousticus

sB   sulcus of Brissaud

sbi   sulcus brevis insulae (short insular sulcus)

sbi I   sulcus brevis insulae, first (first short insular sulcus)

sbi II   sulcus brevis insulae, second (second short insular sulcus)

sbps   subparietal sulcus (splenial sulcus)

ScG   subcentral gyrus

ScaG   subcallosal gyrus

sf   sulcus fragmentosus

SFG   superior frontal gyrus

sfps   superior frontal paramidline sulcus

sfs   superior frontal sulcus

sfs-a   superior frontal sulcus, anterior segment

sfs-p   superior frontal sulcus, posterior segment

sH   sulcus of Heschl

SlG   semilunar gyrus

SmG   supramarginal gyrus

sms   supramarginal sulcus

sos   supra-orbital sulcus

spcs   superior post-central sulcus

SPL   superior parietal lobule

sprs   superior precentral sulcus

sps   superior parietal sulcus

sros   superior rostral sulcus

sSc   sulcus of Schwalbe

STG   superior temporal gyrus

sts   superior temporal sulcus (parallel sulcus)

sts-an   annectant ramus of the superior temporal sulcus

T

tamts   transverse anterior medial temporal sulcus (anterior transverse collateral sulcus)

tcos   transverse collateral sulcus (sulcus collateralis transversus)

Th   thalamus

ti   temporal incisure (incisura temporalis)

TLGP   temporo-limbic gyral passage

tocs   transverse occipital sulcus

tocs-l   transverse occipital sulcus, lateral ramus

tocs-m   transverse occipital sulcus, medial ramus

toi   temporo-occipital incisure

tos   transverse orbital sulcus

tps   temporopolar sulcus

Tr   triangular part of the inferior frontal gyrus (pars triangularis)

ts   triangular sulcus (incisura capitis)

tts   transverse temporal sulcus

U

Un   uncus (uncus hippocampi)

uns   uncal sulcus

MRI Sections

This part of the atlas presents a series of coronal, horizontal, and sagittal sections from the magnetic resonance image (MRI) of one brain transformed into Montreal Neurological Institute (MNI) standard stereotaxic space. The sections are taken at 6 mm intervals in MNI space, except where closer sections are necessary to visualize certain sulci and gyri of the core language regions. All the sulci in the peri-Sylvian region, where the core language areas are located, are identified. Subcortical structures, such as the caudate nucleus, putamen, claustrum, etc, are identified only on some sections to avoid crowding of the images. The level of the coronal (y), horizontal (z) and sagittal (x) sections in MNI standard stereotaxic space is indicated on the top left side of each section and also marked by a black line on the three-dimensional reconstruction of the brain. This three-dimensional reconstruction was created using the image process-ing pipeline Civet (Ad-Dab'bagh et al., 2006). The MRI scan was acquired on a 3T Philips Gyroscan superconducting magnet system. The T1 anatomical data were then tranformed into MNI standard ste-reotaxic space using a 12 parameter linear registration to the ICBM152 generation VI average symmetric brain (Grabner et al., 2006). The T1 scan was then re-sampled on a $1mm^3$ (voxel size) isotropic grid. The thickness of the coronal, horizontal, and sagittal slices is 1mm. The letters L and R are used to indicate the left and right hemispheres. The y values refer to the antero-posterior distance relative to the origin (positive values = anterior to the origin, negative values = posterior to the origin). The x values refer to the medial-to-lateral distance relative to the midline (negative values = left hemisphere). The z values refer to the superior-inferior distance from the origin (positive values = superior to the origin, negative values = inferior to the origin).

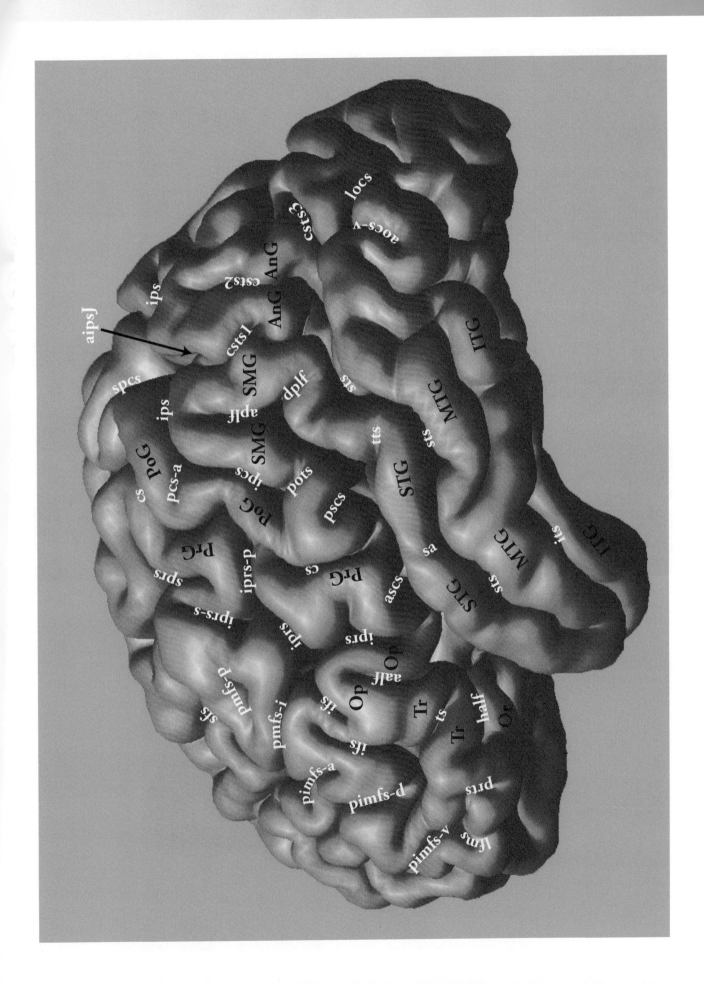

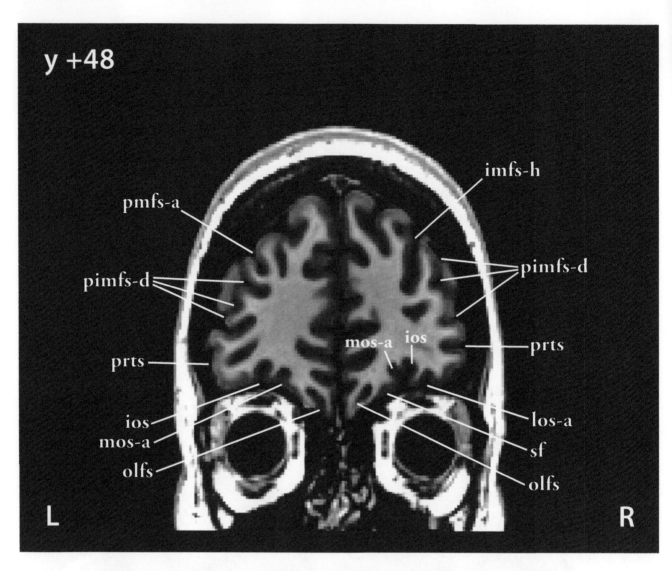

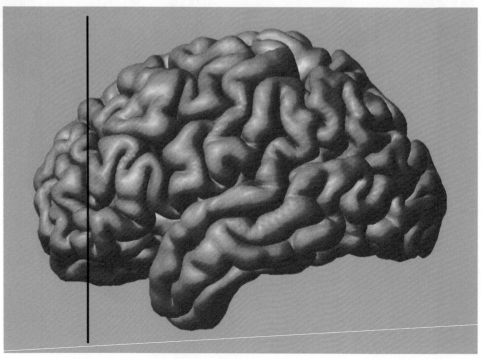

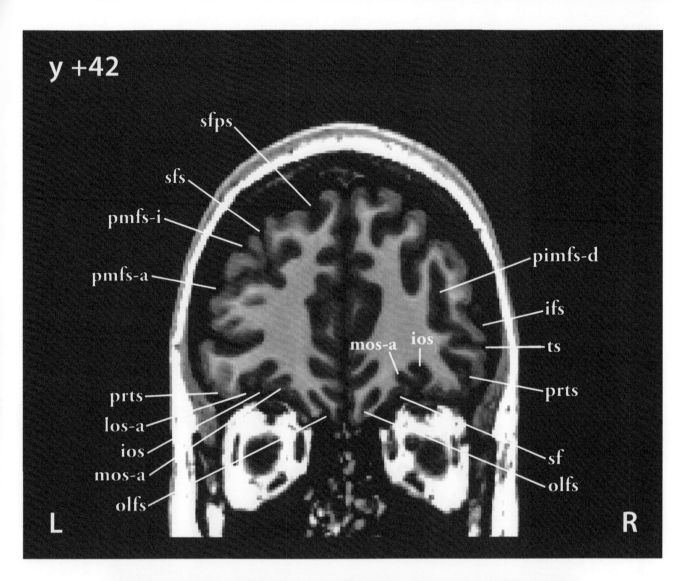

y +42

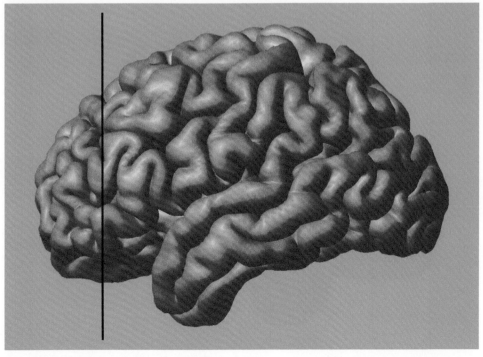

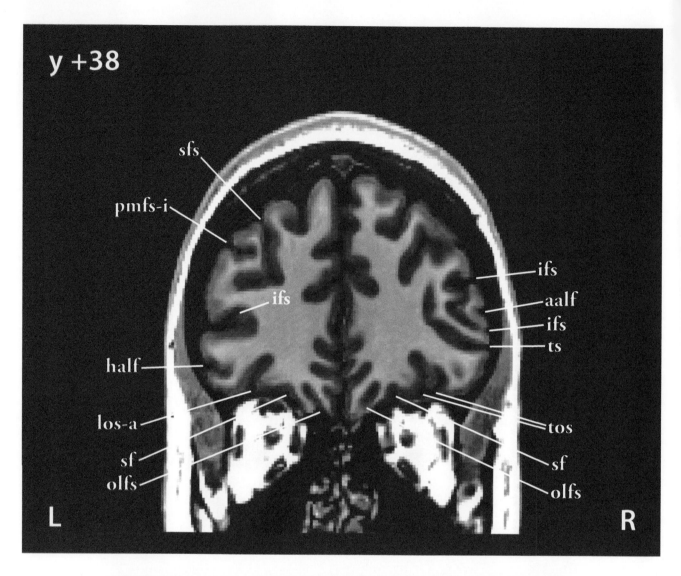

y +38

sfs

pmfs-i

ifs

ifs
aalf
ifs
ts

half

los-a

tos

sf

sf
olfs

olfs

L   R

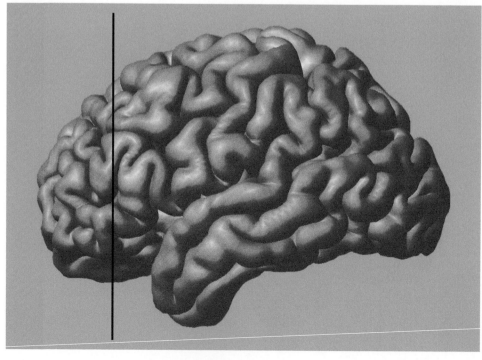

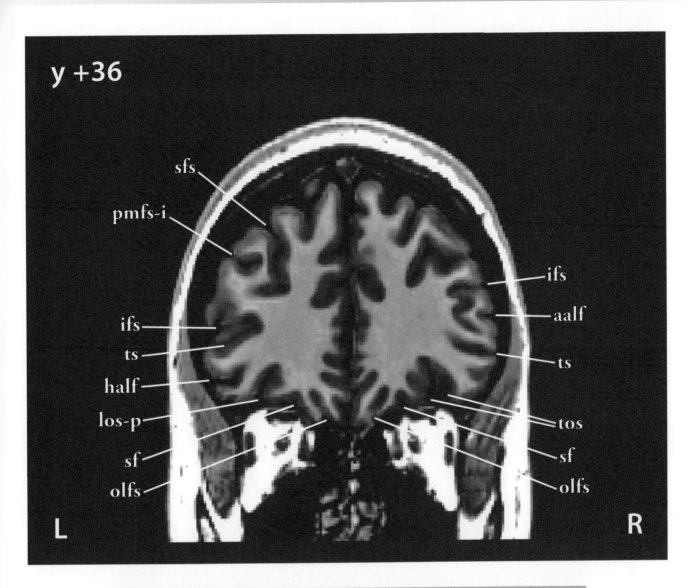

y +36

sfs
pmfs-i
ifs
ts
half
los-p
sf
olfs
ifs
aalf
ts
tos
sf
olfs
L
R

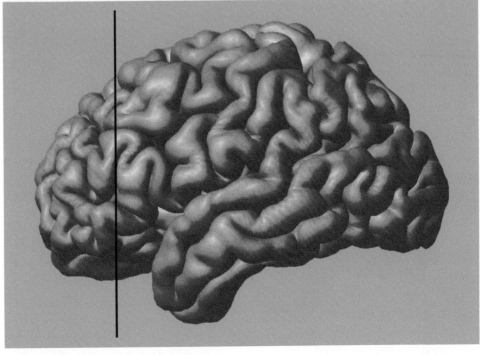

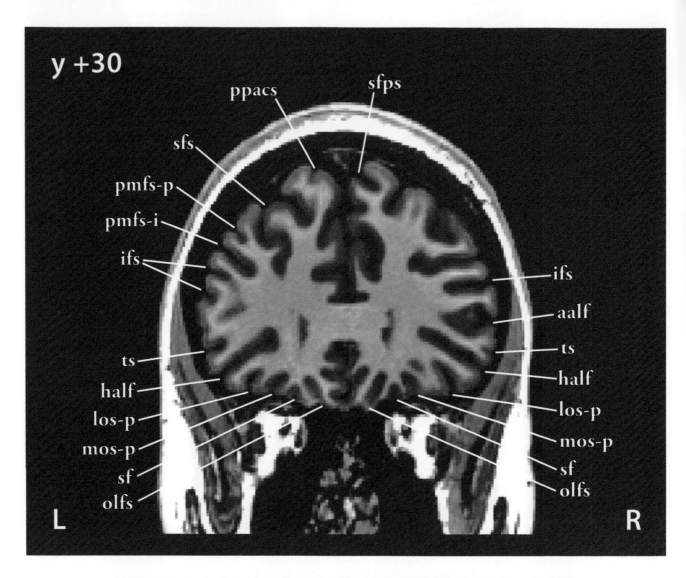

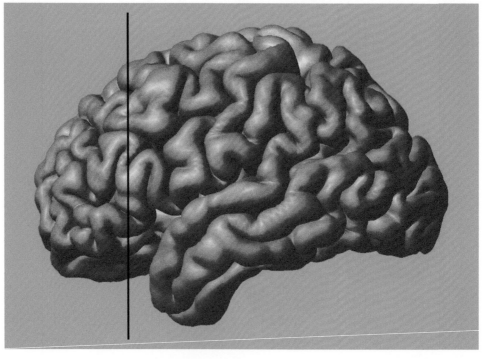

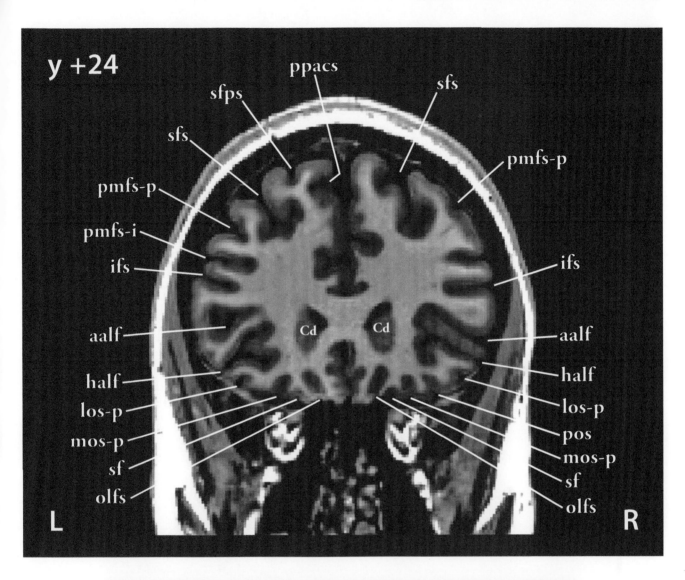

y +24

ppacs

sfps

sfs

sfs

pmfs-p

pmfs-p

pmfs-i

ifs

ifs

aalf

aalf

half

half

los-p

los-p

mos-p

pos

sf

mos-p

olfs

sf

olfs

Cd   Cd

L

R

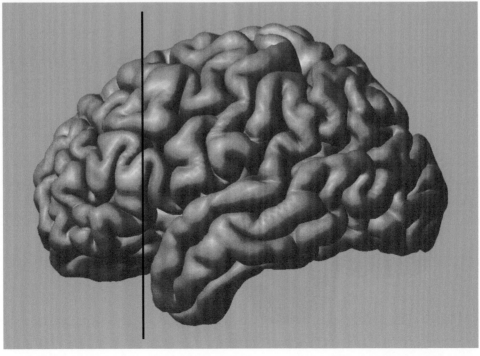

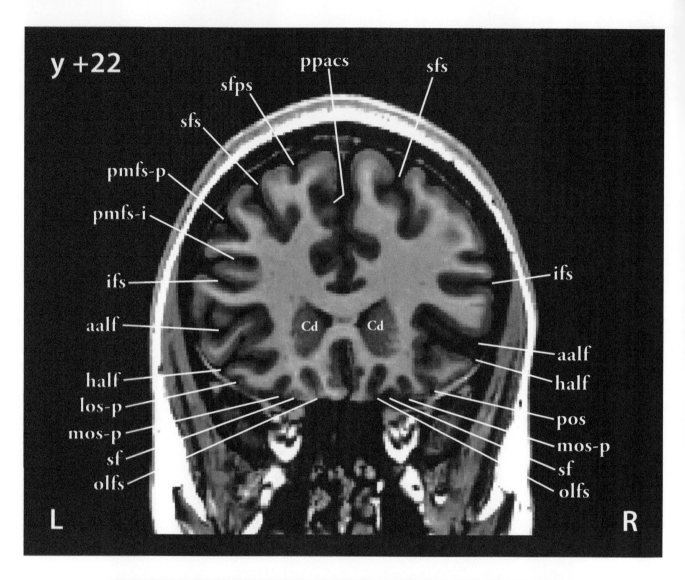

y +22

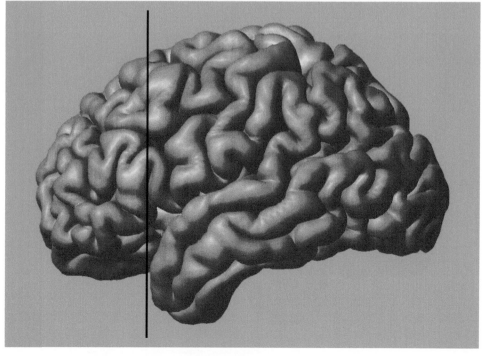

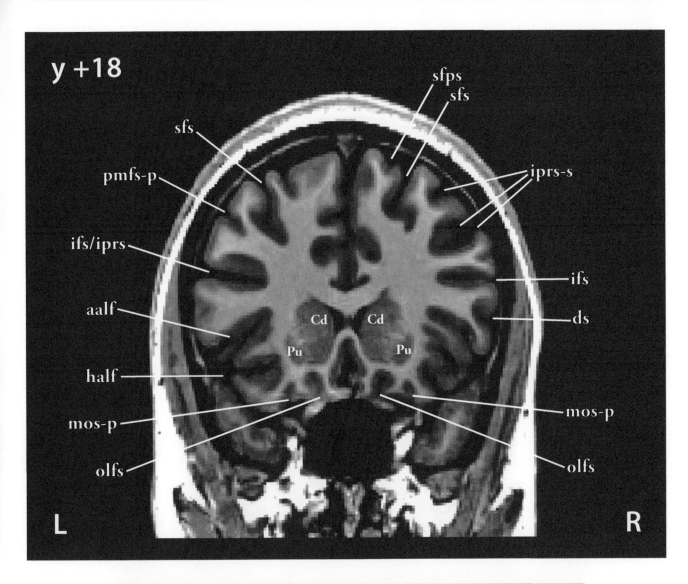

y +18

sfps
sfs
sfs
pmfs-p
iprs-s
ifs/iprs
ifs
aalf
ds
Cd   Cd
Pu   Pu
half
mos-p
mos-p
olfs
olfs

L   R

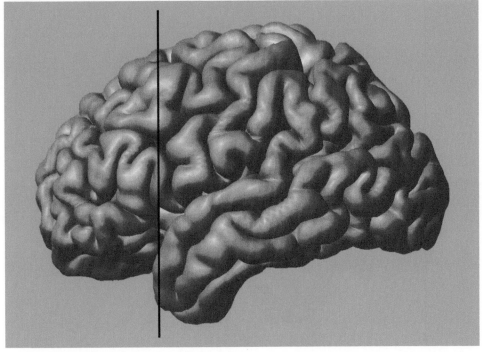

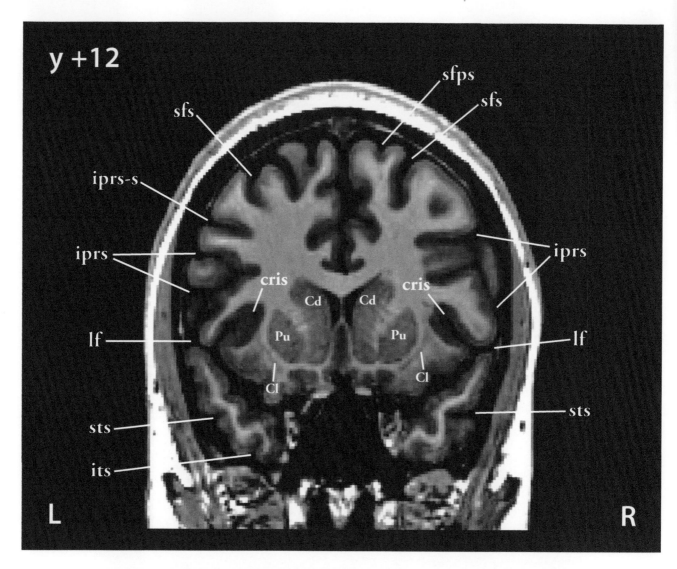

y +12

sfps
sfs
sfs
iprs-s
iprs
iprs
cris
cris
Cd
Cd
Pu
Pu
lf
lf
Cl
Cl
sts
sts
its

L
R

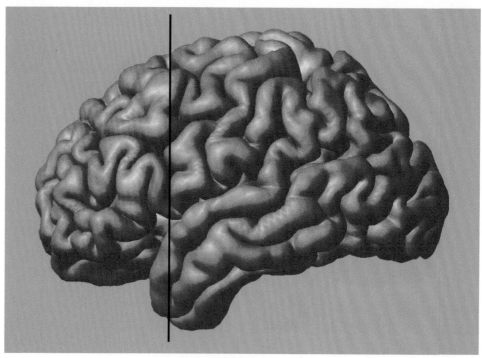

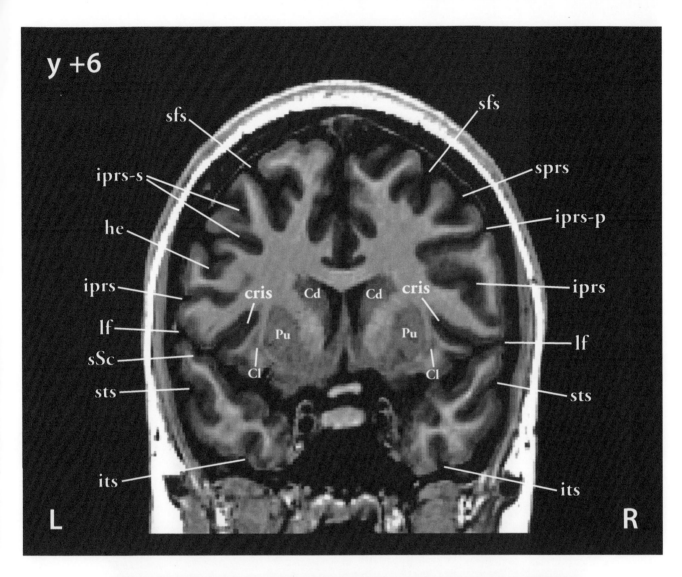

y +6

sfs

sfs

iprs-s

sprs

he

iprs-p

iprs

cris

Cd

Cd

cris

iprs

lf

Pu

Pu

lf

sSc

Cl

Cl

sts

sts

its

its

L

R

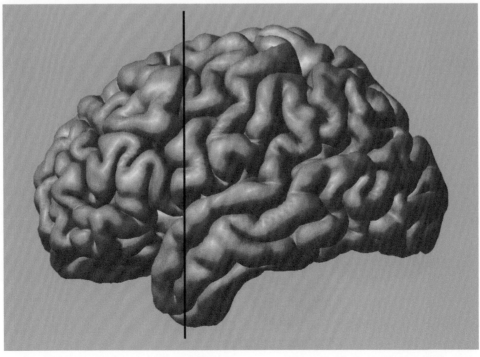

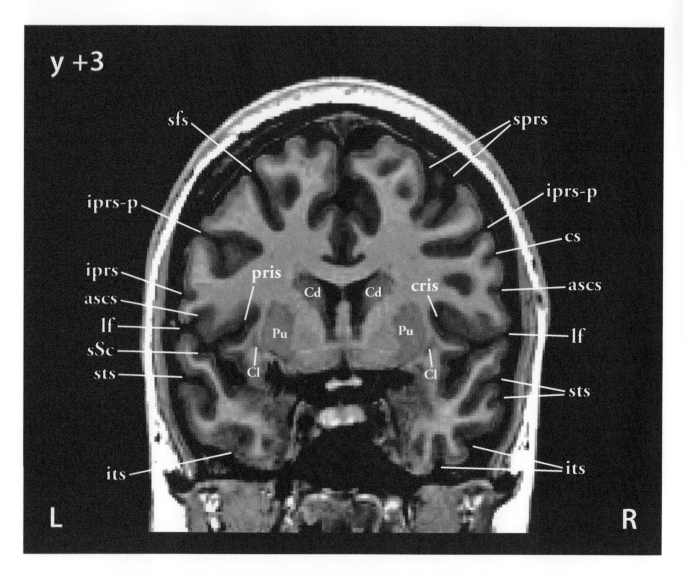

y +3

sfs

sprs

iprs-p

iprs-p

cs

iprs

pris

Cd   Cd

cris

ascs

ascs

lf

Pu

Pu

lf

sSc

Cl

Cl

sts

sts

its

its

L

R

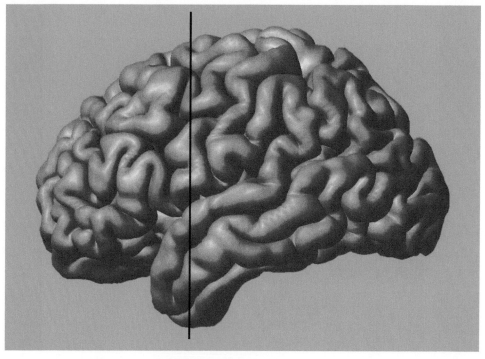

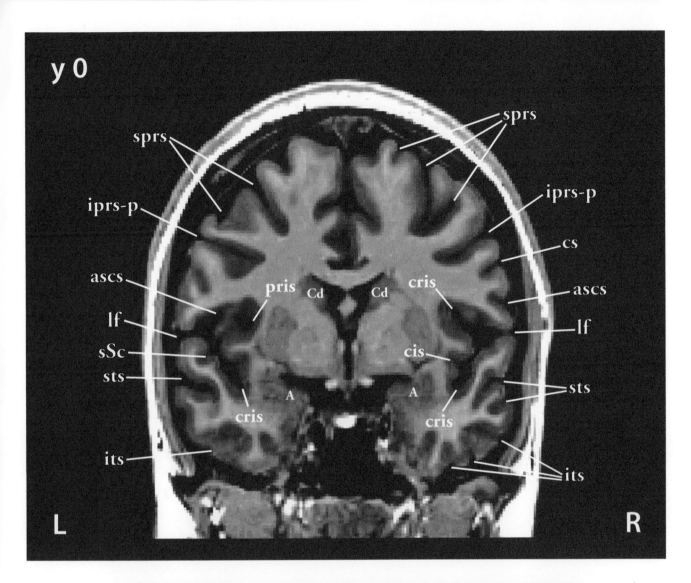

y 0

L   R

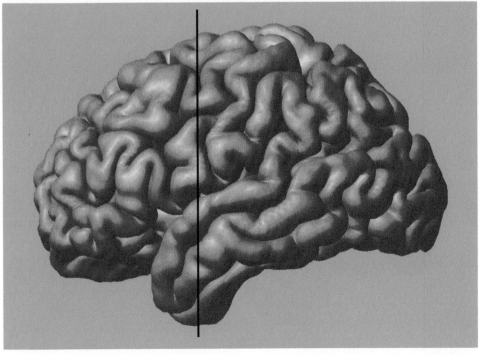

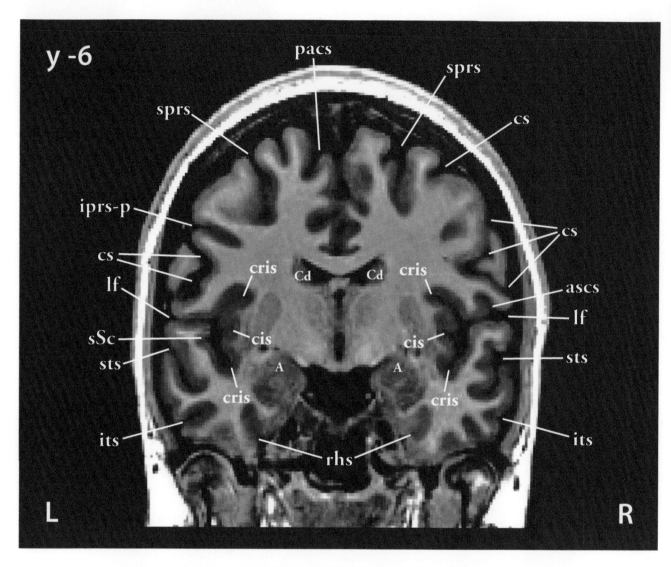

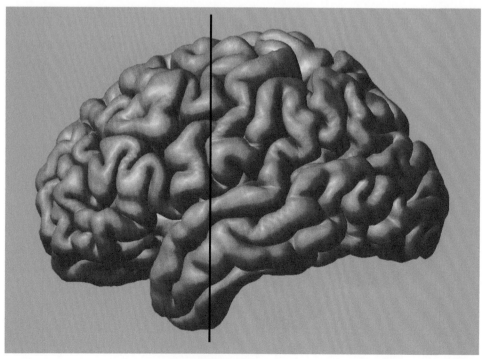

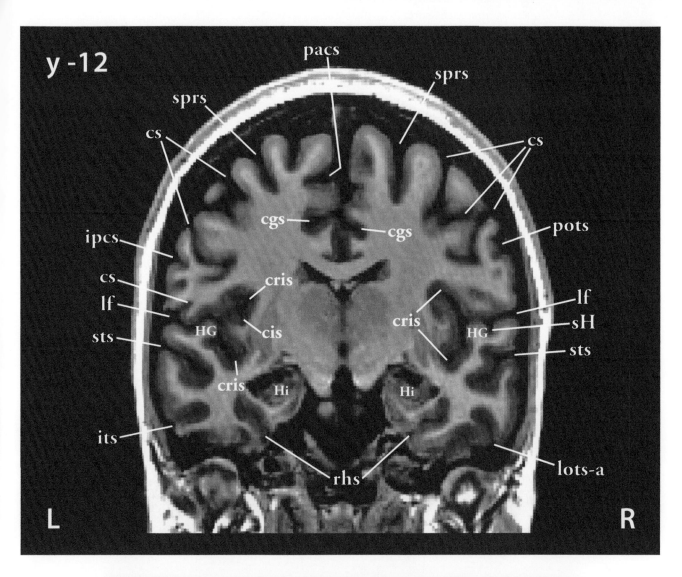

y -12

pacs

sprs

sprs

cs

cs

cgs

cgs

pots

ipcs

cris

cs

cris

lf

lf

HG

sH

sts

cis

HG

sts

cris

Hi

Hi

its

rhs

lots-a

L

R

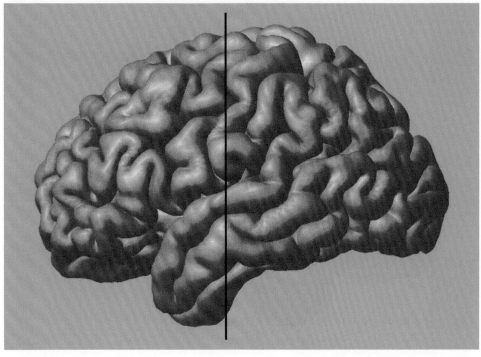

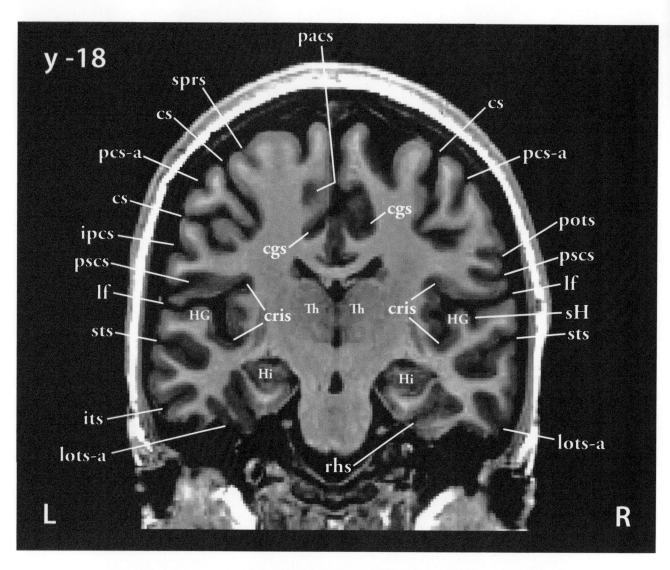

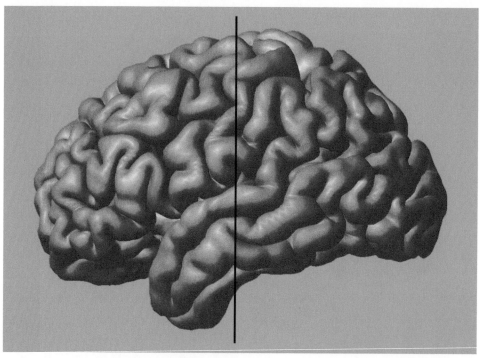

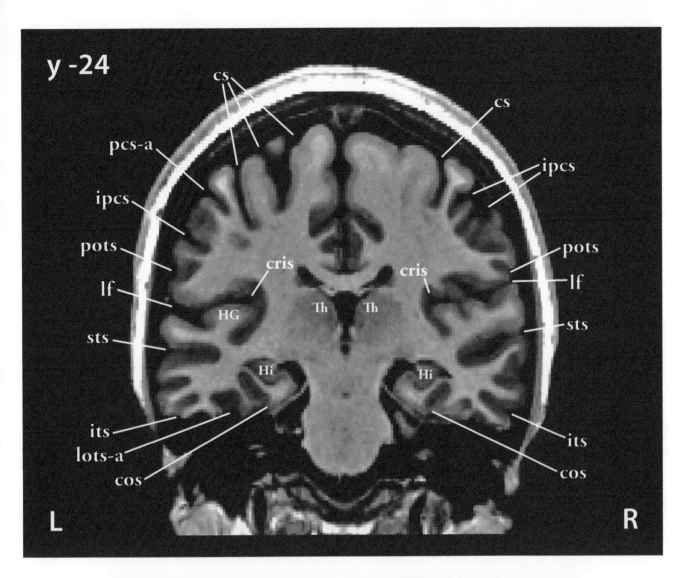

y -24

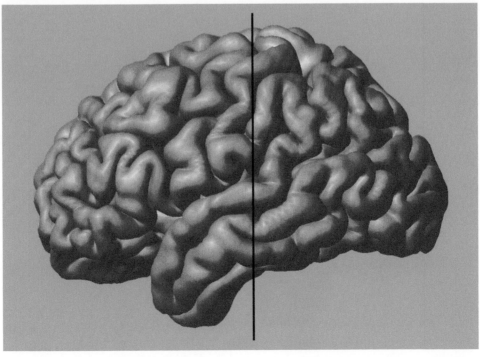

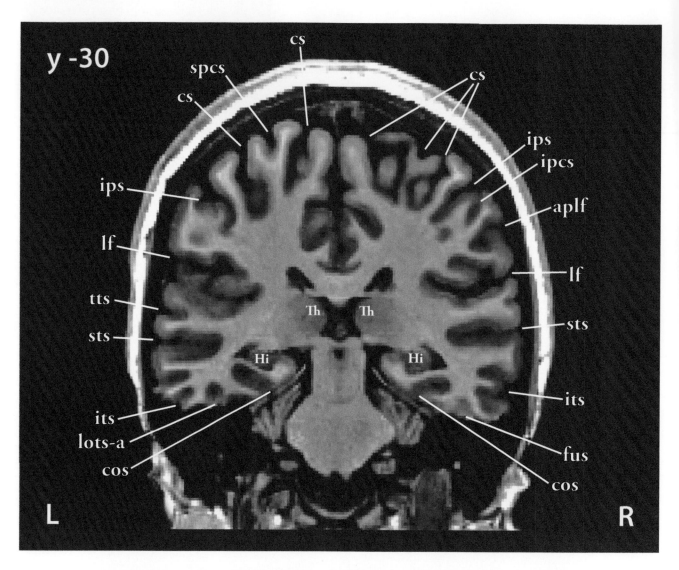

y -30

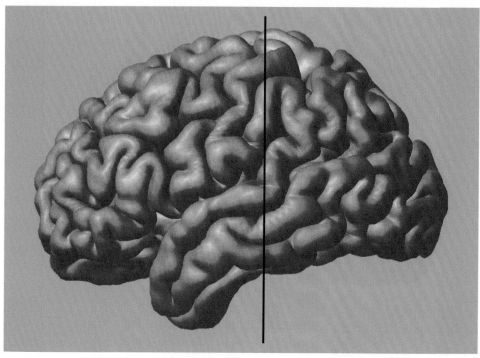

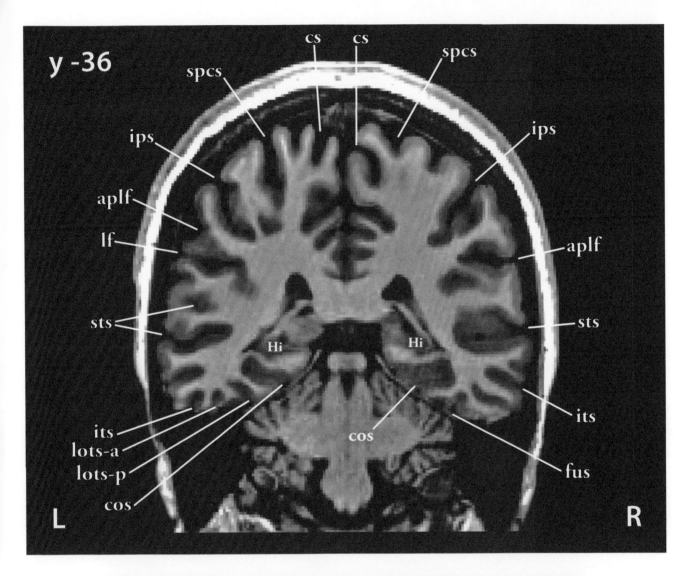

y -36

cs cs
spcs spcs
ips ips
aplf aplf
lf
sts sts
Hi Hi
its
lots-a
lots-p
cos
cos
fus
its

L R

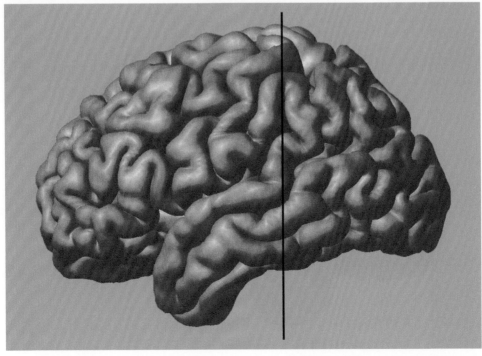

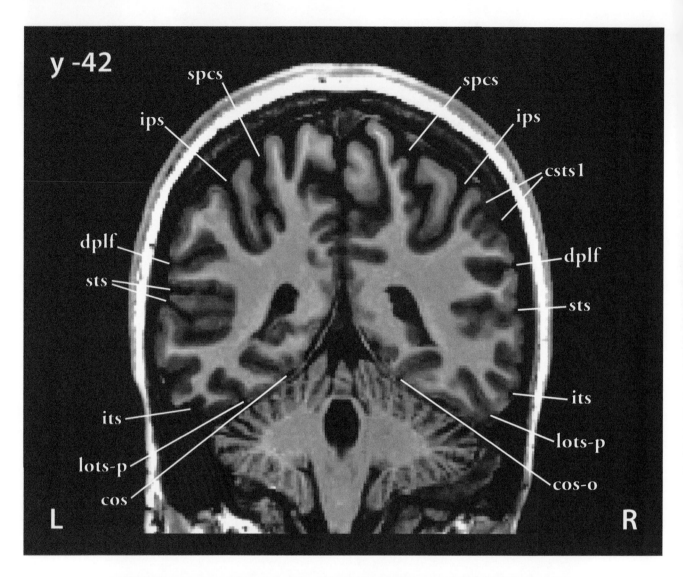

y -42

spcs spcs
ips ips
csts 1
dplf dplf
sts sts
dplf
its its
lots-p lots-p
cos cos-o
L R

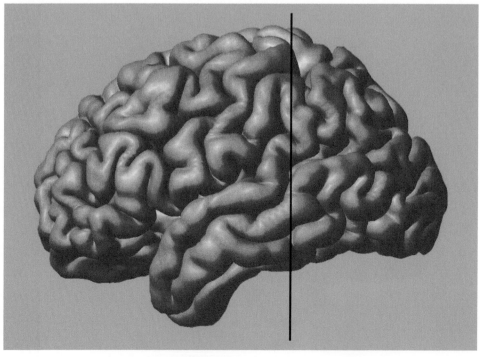

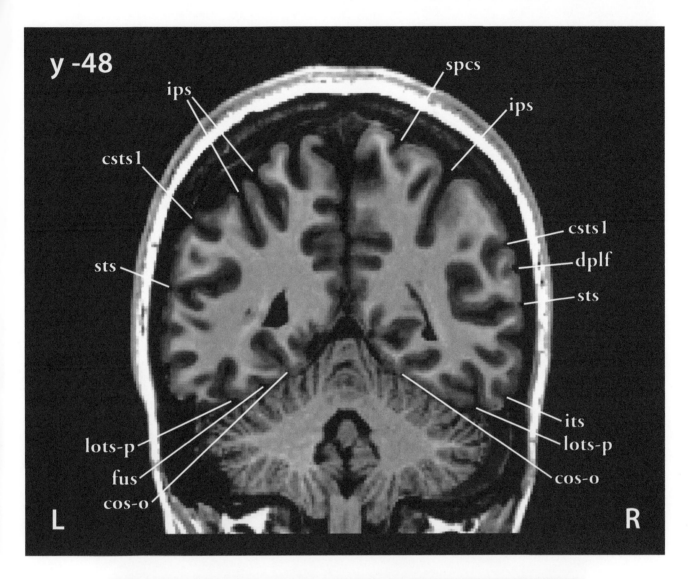

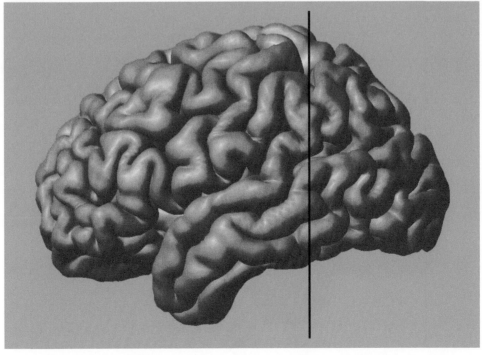

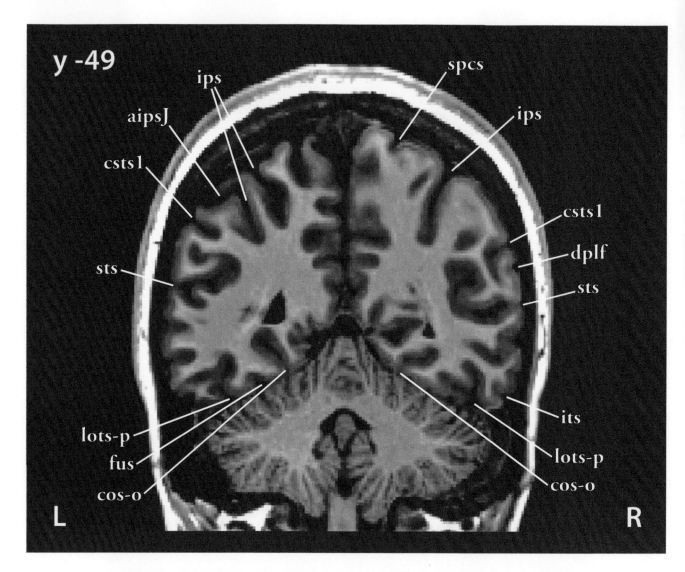

y -49

ips
spcs
aipsJ
ips
csts1
csts1
dplf
sts
sts
its
lots-p
lots-p
fus
cos-o
cos-o

L   R

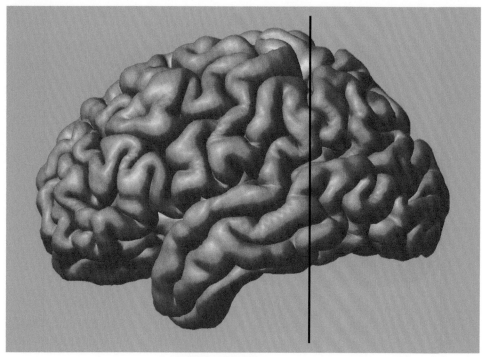

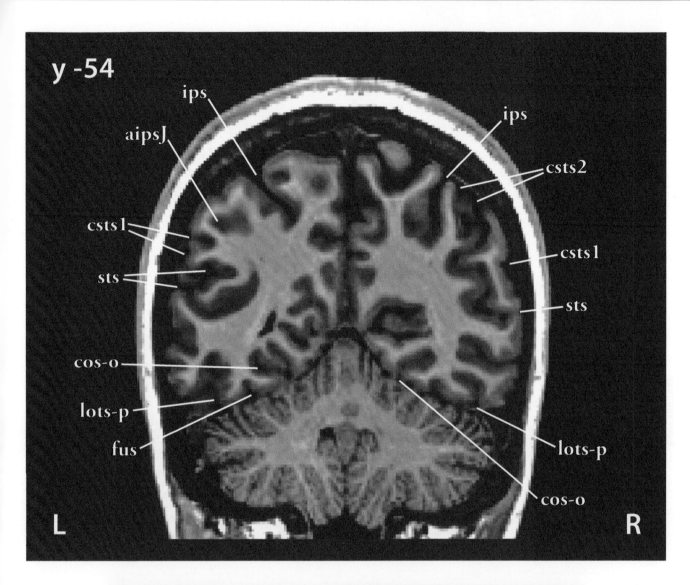

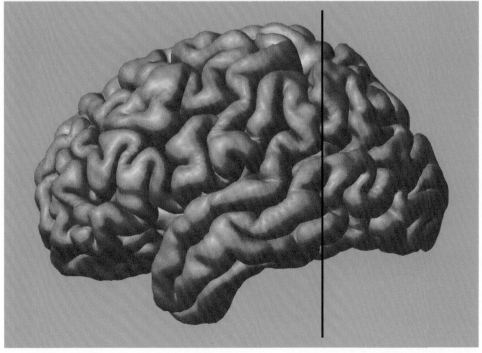

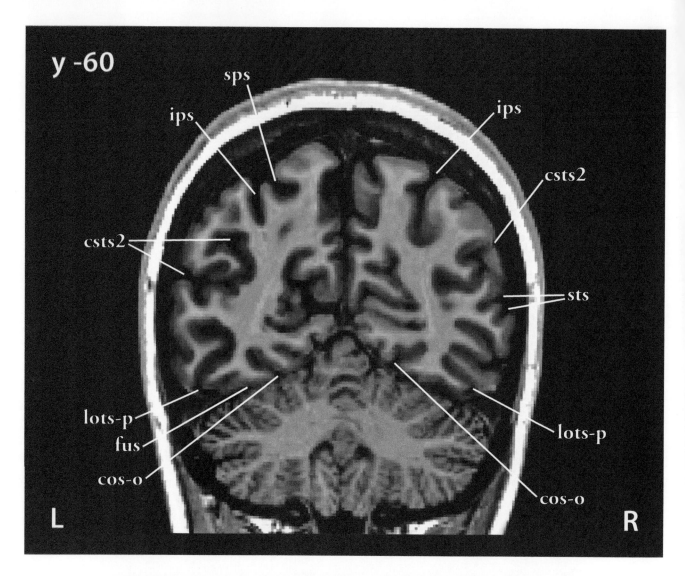

y -60

sps
ips
ips
csts2
csts2
sts
lots-p
fus
cos-o
lots-p
cos-o
L
R

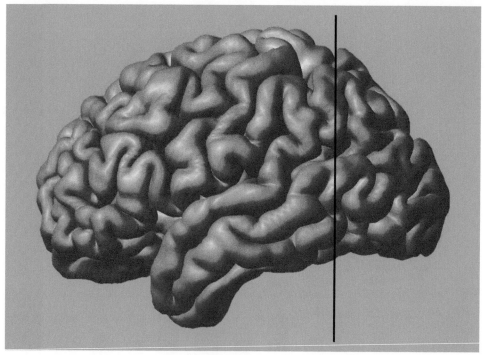

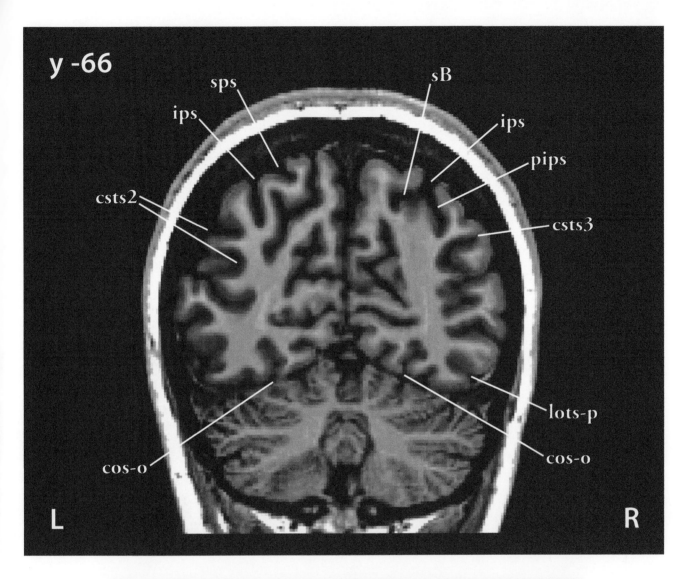

y -66

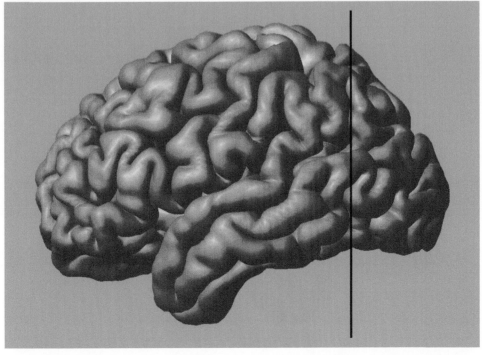

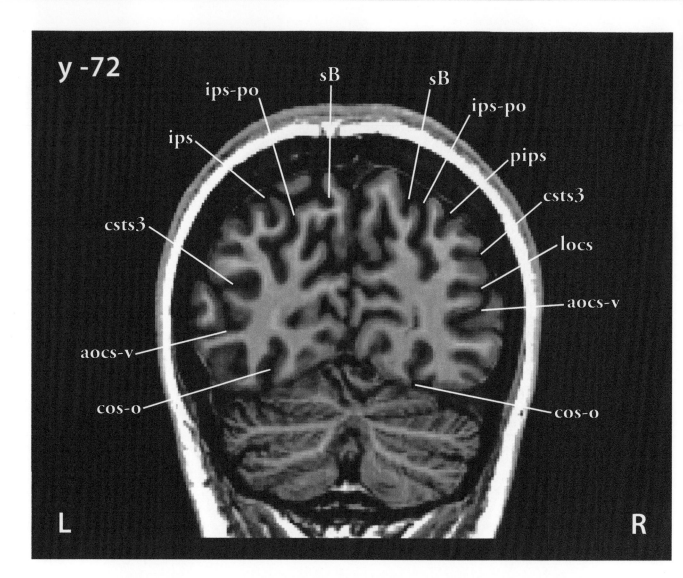

y -72

ips-po
sB
sB
ips-po
ips
pips
csts3
csts3
locs
aocs-v
cos-o
aocs-v
cos-o

L

R

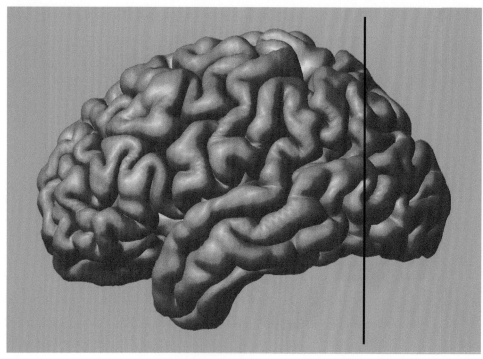

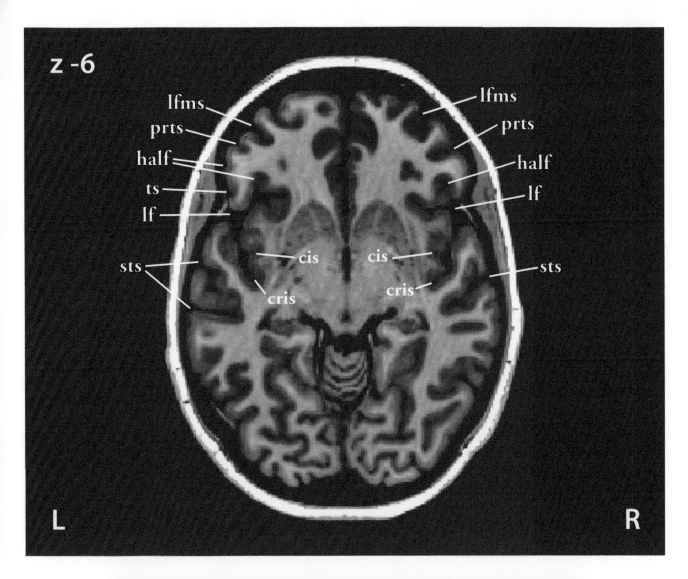

z -6

lfms
prts
half
ts
lf
sts
cis
cris
cis
crris
lfms
prts
half
lf
sts

L
R

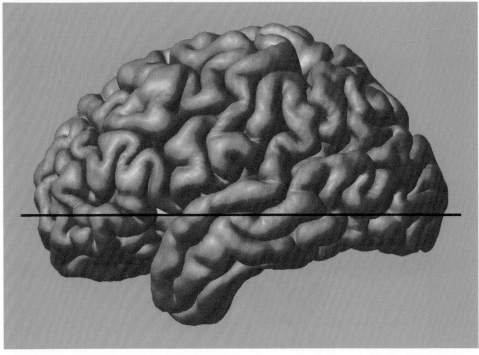

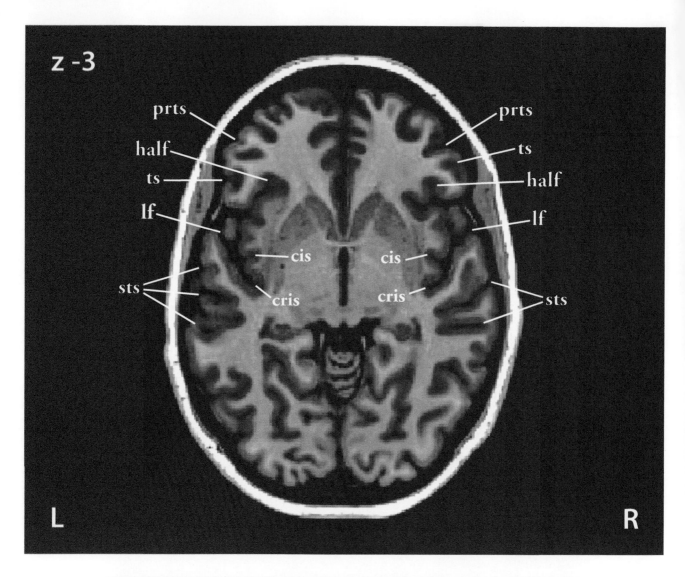

z -3

prts
half
ts
lf
cis
sts
cris

prts
ts
half
lf
cis
cris
sts

L

R

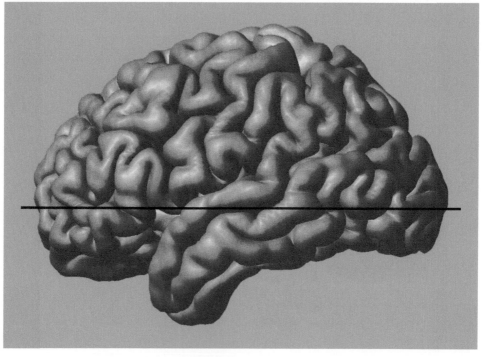

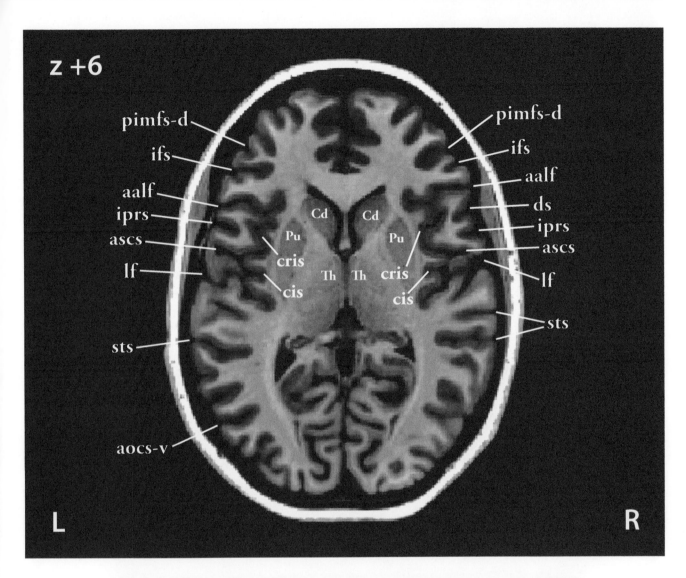

z +6

pimfs-d
ifs
aalf
iprs
ascs
lf
sts
aocs-v

pimfs-d
ifs
aalf
ds
iprs
ascs
lf
sts

Cd Cd
Pu Pu
cris cris
Th Th
cis cis

L

R

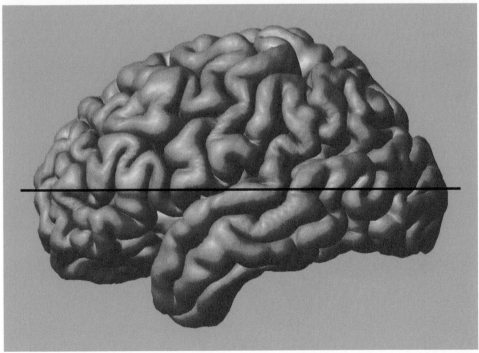

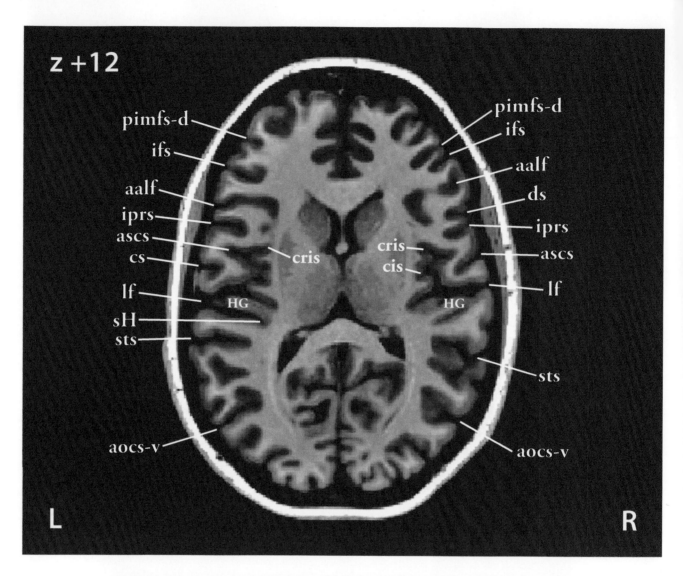

z +12

pimfs-d
ifs
aalf
iprs
ascs
cs
lf
sH
sts

pimfs-d
ifs
aalf
ds
iprs
ascs
lf

cris
cis

cris

HG

HG

sts

aocs-v

aocs-v

L

R

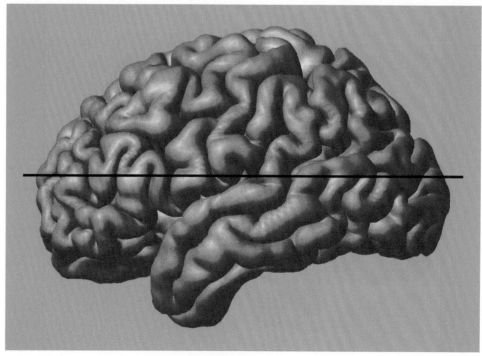

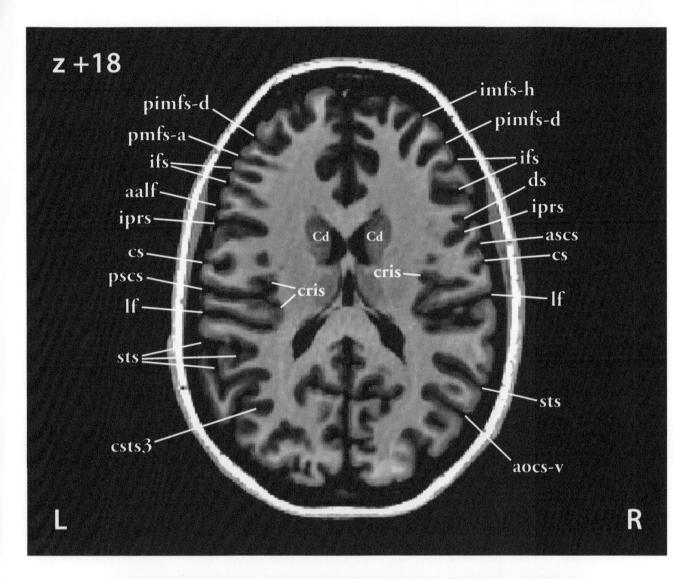

z +18

pimfs-d
pmfs-a
ifs
aalf
iprs
cs
pscs
lf
sts
csts3

imfs-h
pimfs-d
ifs
ds
iprs
ascs
cs
lf
sts
aocs-v

cris
cris

Cd   Cd

L   R

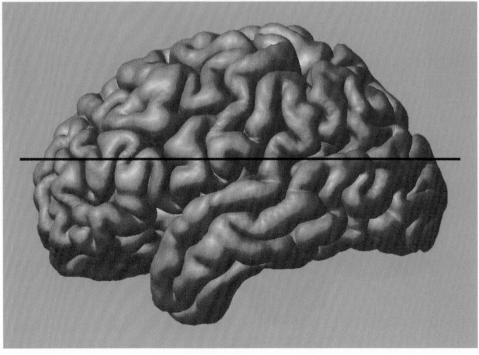

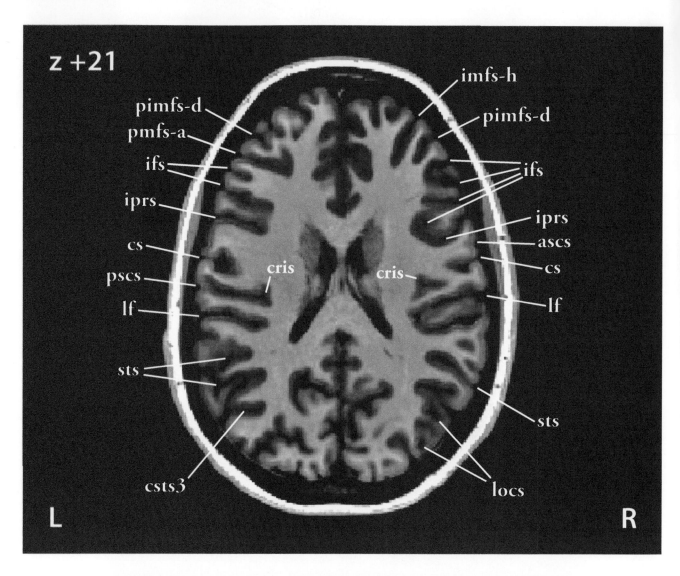

z +21

imfs-h

pimfs-d

pimfs-d

pmfs-a

ifs

ifs

iprs

iprs

cs

ascs

cris

cris

cs

pscs

lf

lf

sts

sts

csts3

locs

L

R

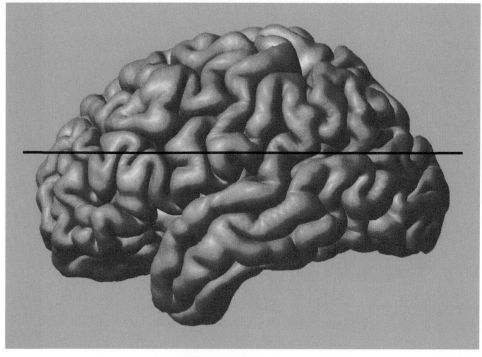

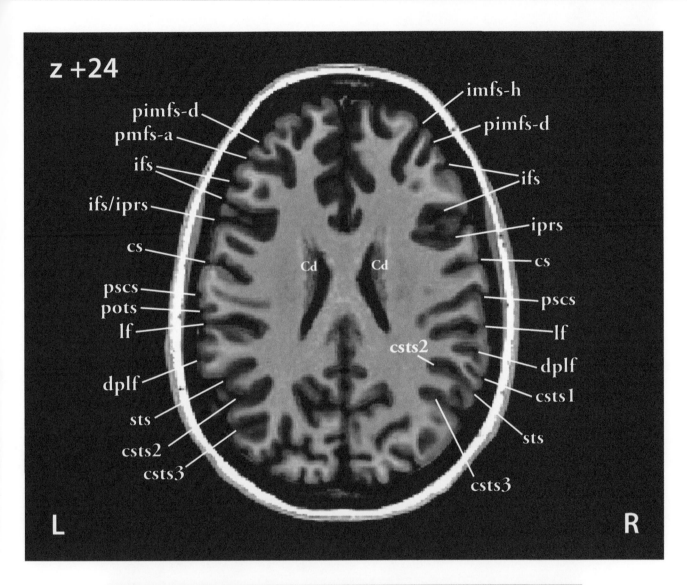

z +24

imfs-h
pimfs-d
pimfs-d
pmfs-a
ifs
ifs
ifs/iprs
iprs
cs
cs
pscs
pscs
pots
lf
lf
dplf
dplf
sts
csts1
csts2
csts2
sts
csts3
csts3

Cd Cd

L R

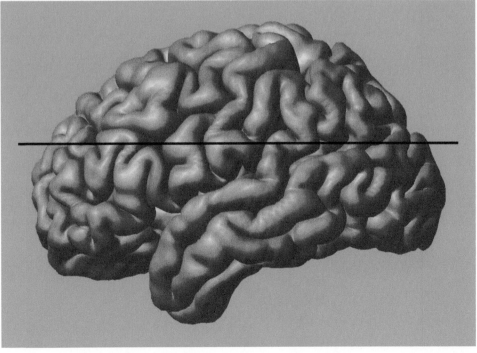

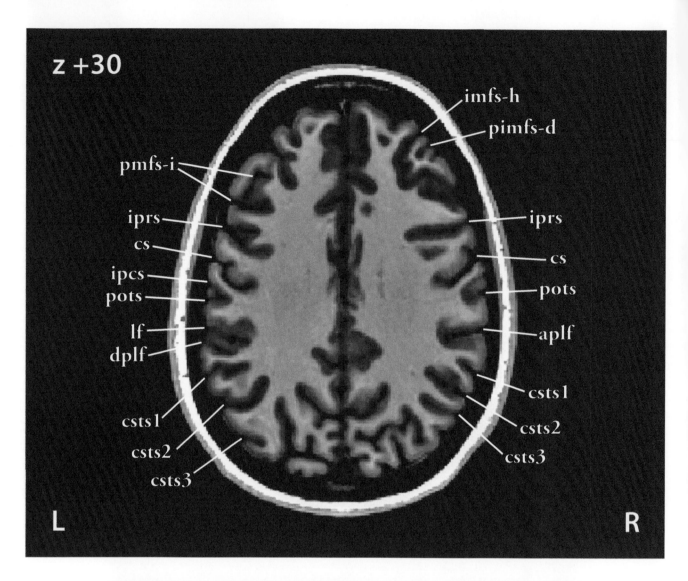

z +30

imfs-h
pimfs-d
pmfs-i
iprs
cs
ipcs
pots
lf
dplf
csts1
csts2
csts3

iprs
cs
pots
aplf
csts1
csts2
csts3

L   R

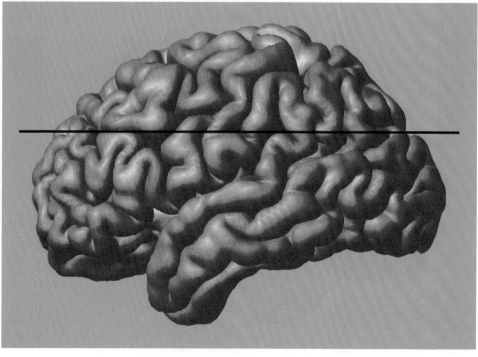

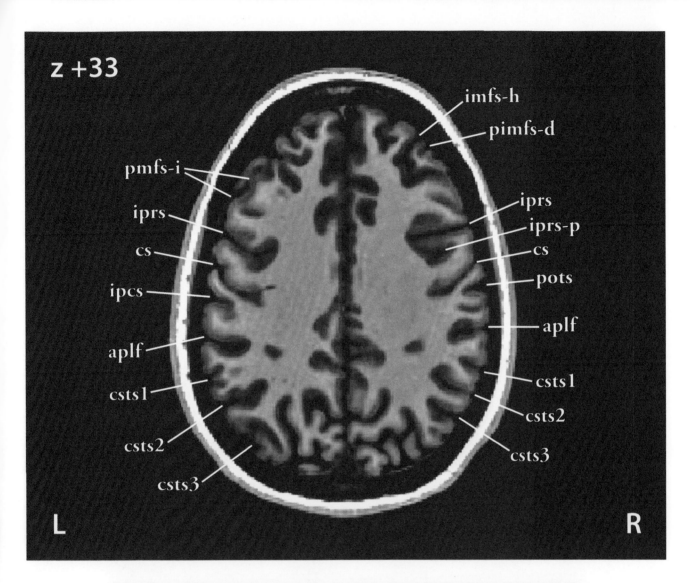

z +33

imfs-h
pimfs-d
pmfs-i
iprs
iprs
iprs-p
cs
cs
ipcs
pots
aplf
aplf
csts1
csts1
csts2
csts2
csts3
csts3

L R

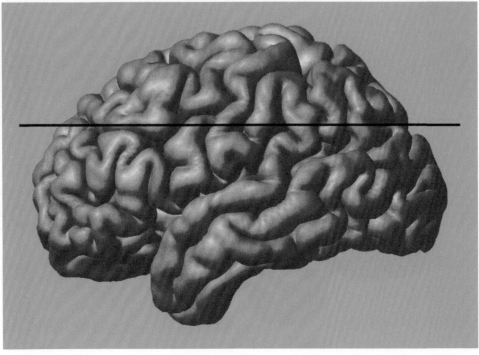

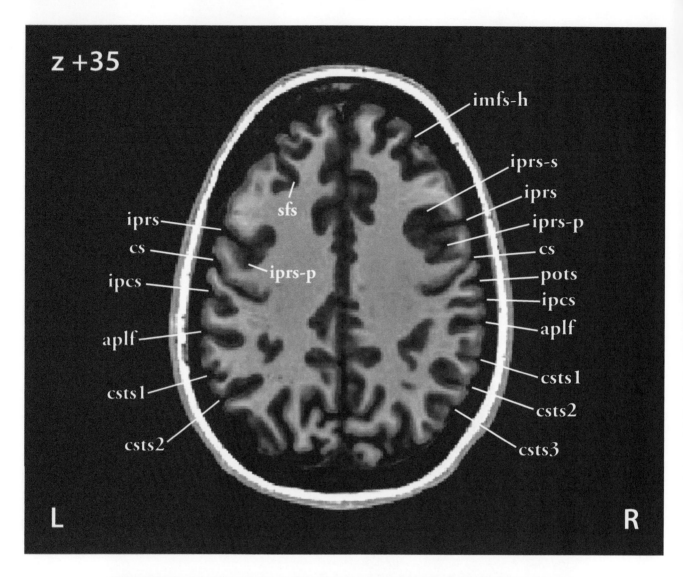

z +35

imfs-h

iprs-s

iprs

iprs-p

cs

pots

ipcs

aplf

csts 1

csts2

csts3

sfs

iprs

cs

ipcs

aplf

csts 1

csts2

iprs-p

L   R

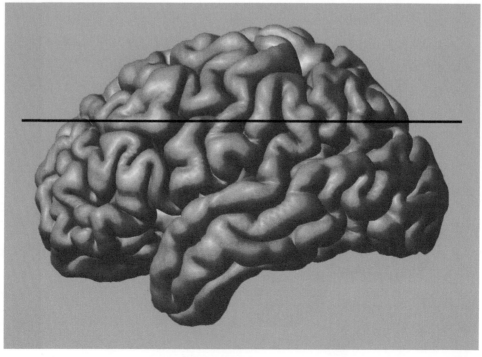

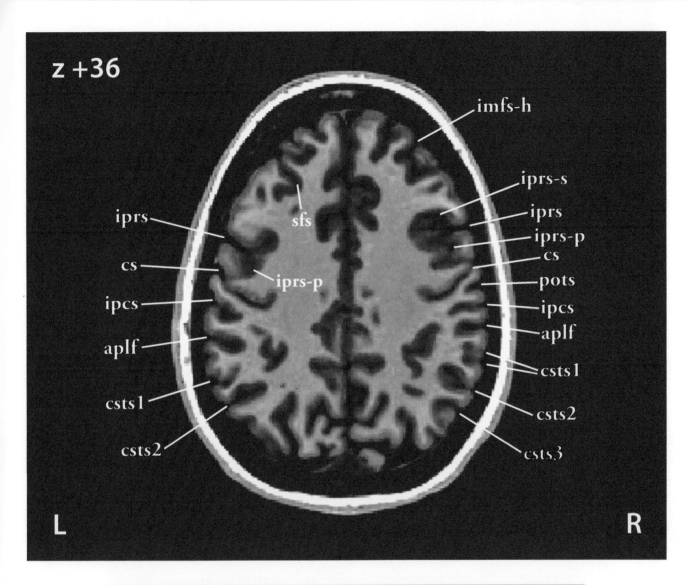

z +36

imfs-h

iprs-s

iprs

iprs

iprs-p

sfs

cs

cs

iprs-p

pots

ipcs

ipcs

aplf

aplf

csts1

csts1

csts2

csts2

csts3

L   R

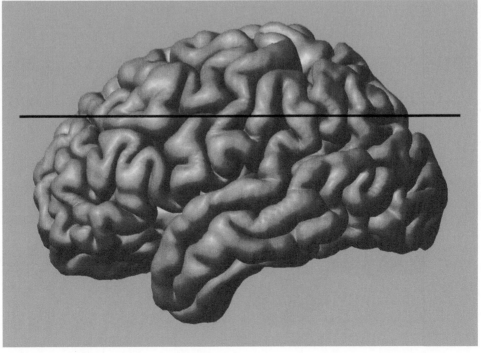

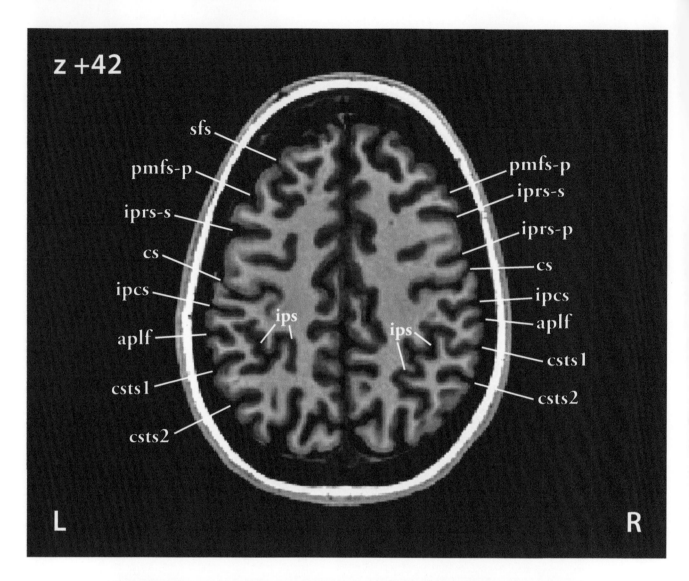

z +42

sfs
pmfs-p
iprs-s
cs
ipcs
aplf
csts1
csts2
ips

pmfs-p
iprs-s
iprs-p
cs
ipcs
aplf
csts1
csts2
ips

L

R

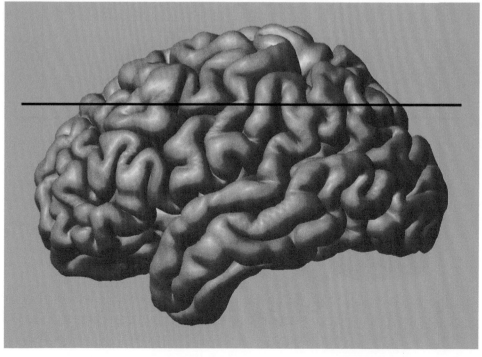

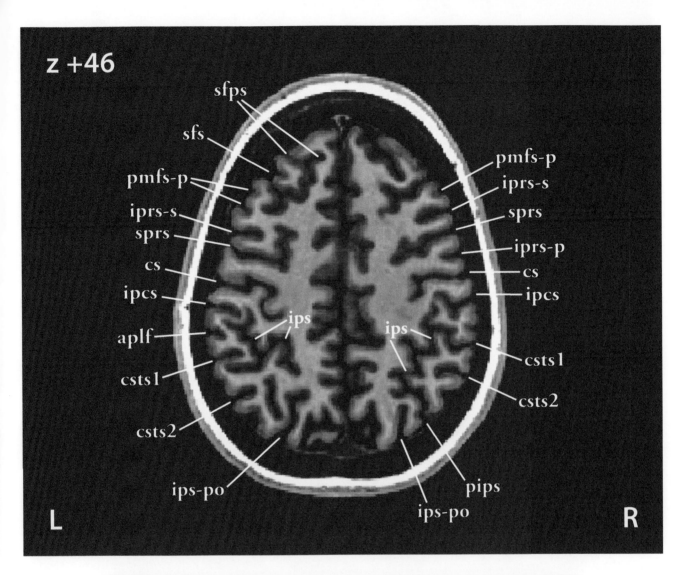

z +46

sfps
sfs
pmfs-p
iprs-s
sprs
cs
ipcs
aplf
csts1
csts2
ips-po

pmfs-p
iprs-s
sprs
iprs-p
cs
ipcs
ips
csts1
csts2
pips
ips-po

L

R

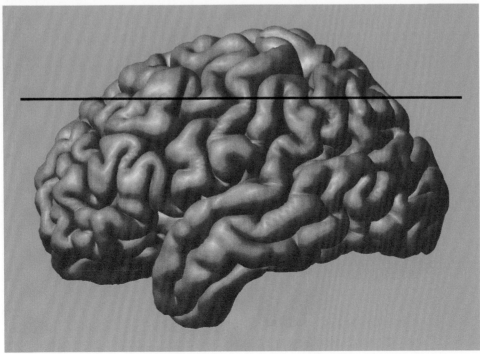

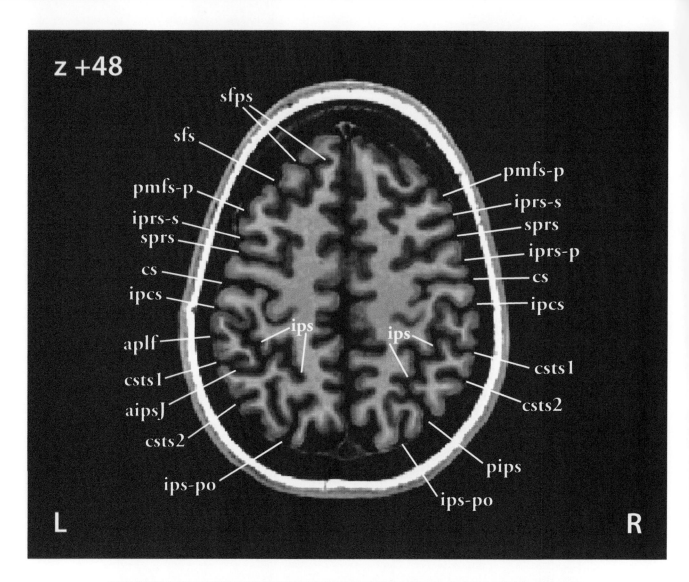

z +48

sfps
sfs
pmfs-p
iprs-s
sprs
cs
ipcs
aplf
csts1
aipsJ
csts2
ips-po

ips
ips

pmfs-p
iprs-s
sprs
iprs-p
cs
ipcs
csts1
csts2
pips
ips-po

L
R

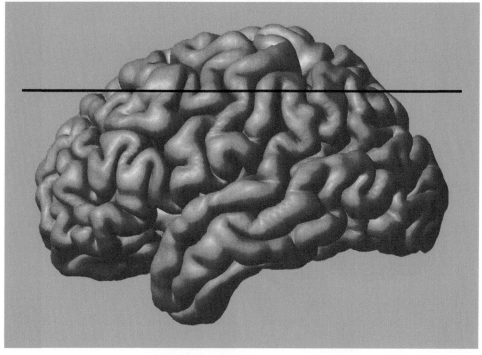

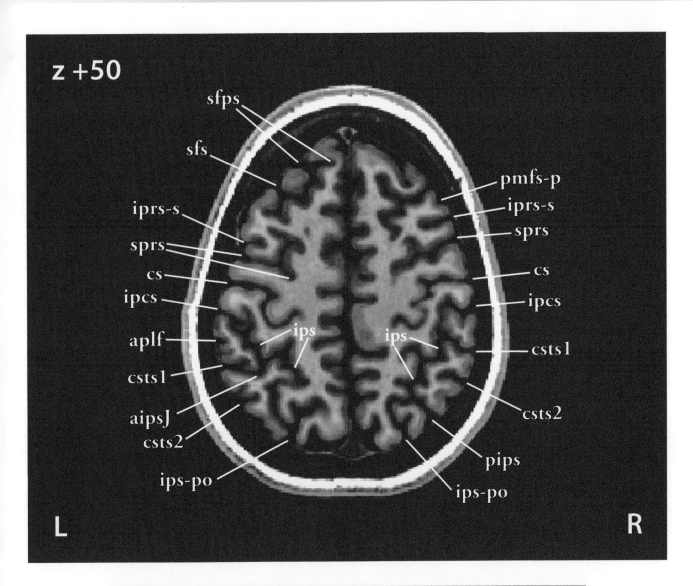

z +50

sfps
sfs
pmfs-p
iprs-s
iprs-s
sprs
sprs
cs
cs
ipcs
ipcs
aplf
ips
ips
csts1
csts1
csts2
aipsJ
csts2
pips
ips-po
ips-po

L
R

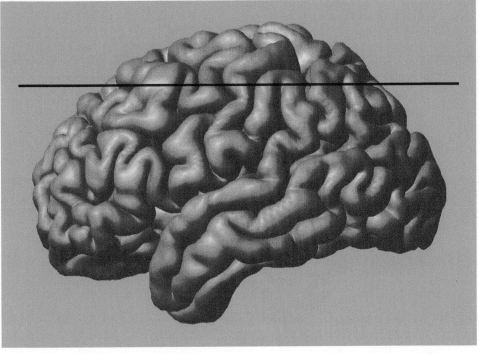

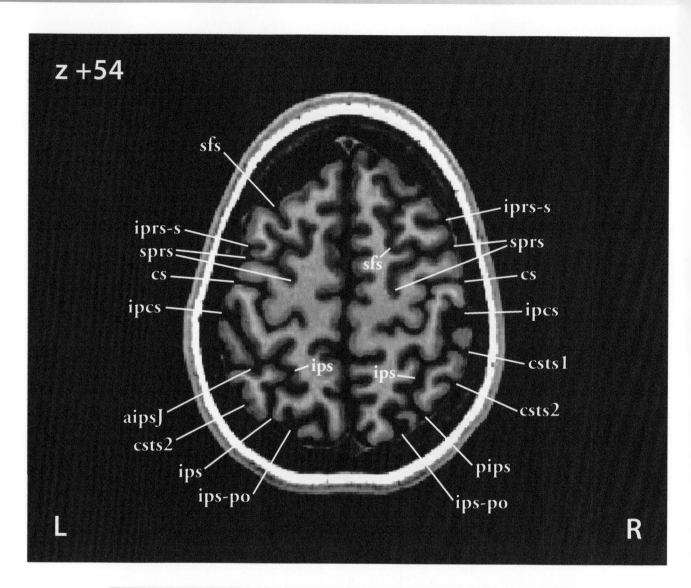

z +54

sfs

iprs-s

sprs

cs

ipcs

sfs

iprs-s

sprs

cs

ipcs

csts 1

ips

ips

aipsJ

csts2

ips

ips-po

pips

ips-po

csts2

L

R

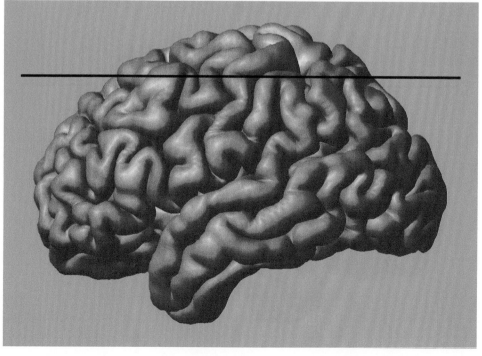

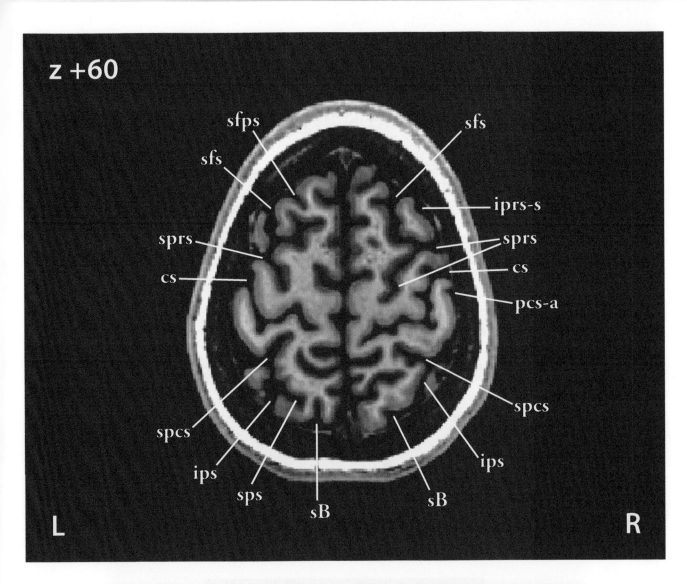

z +60

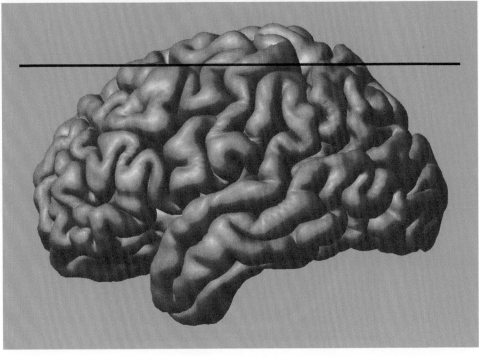

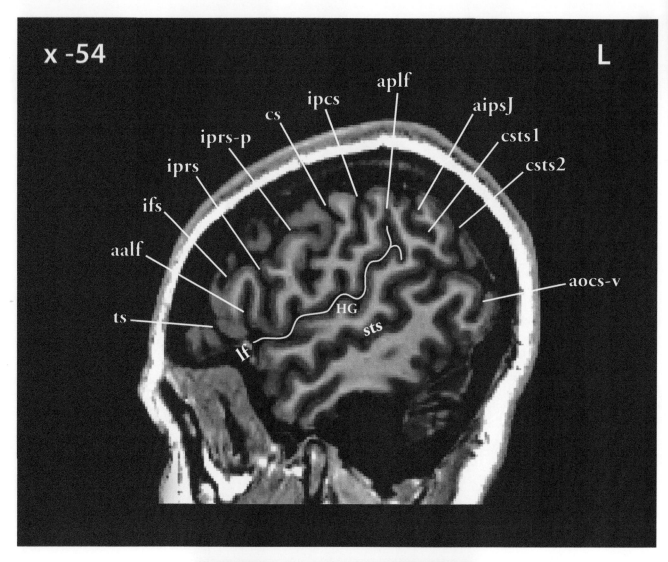

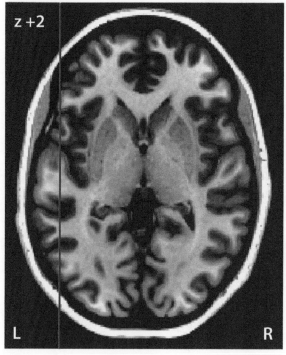

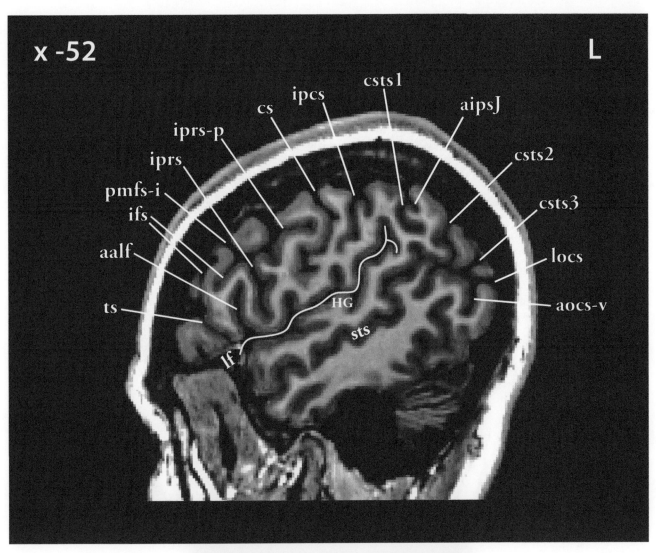

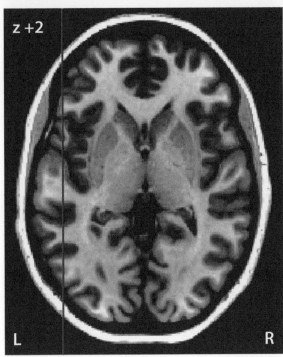

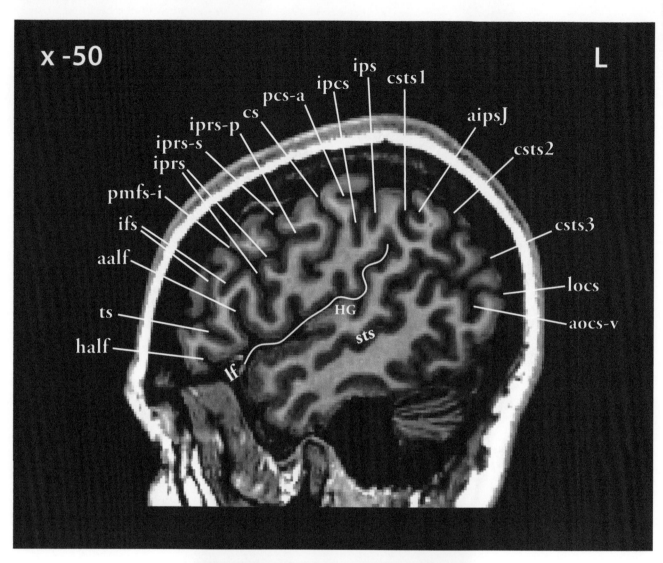

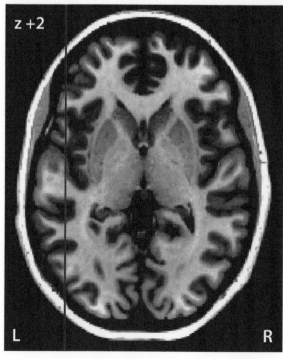

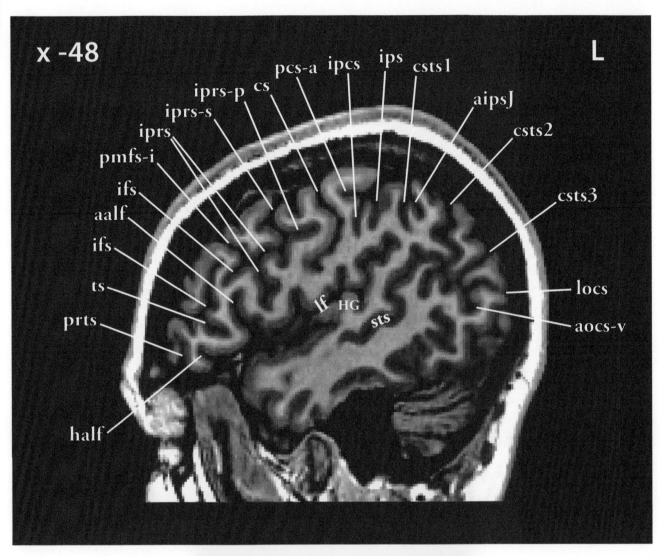

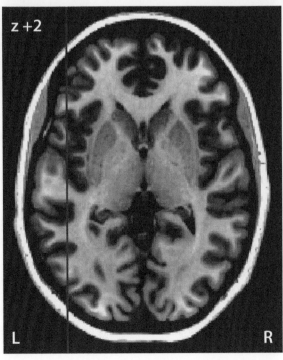

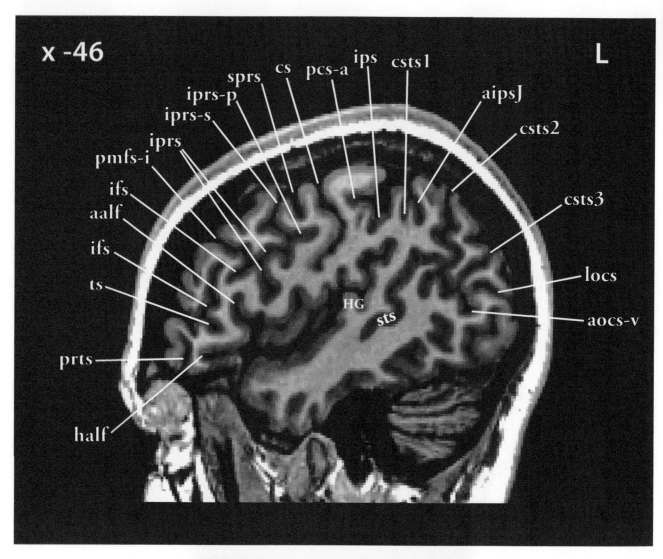

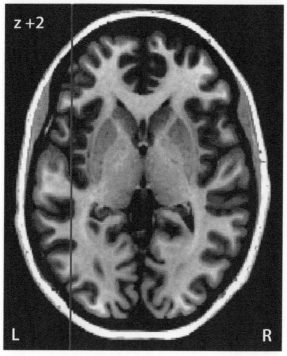

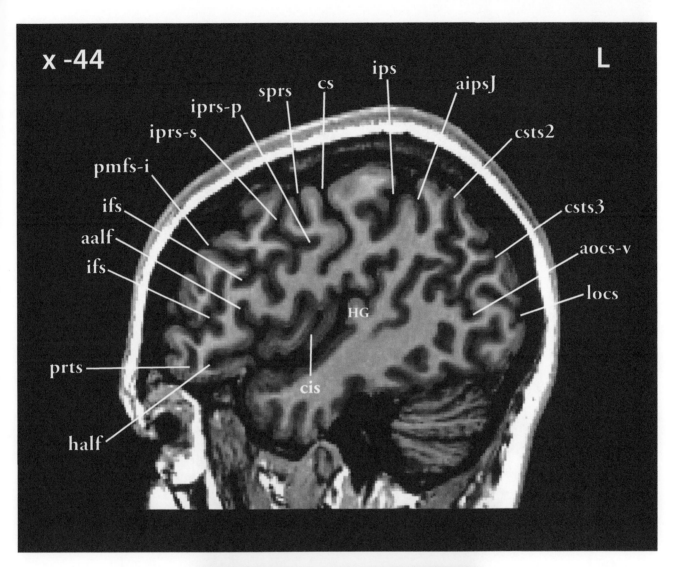

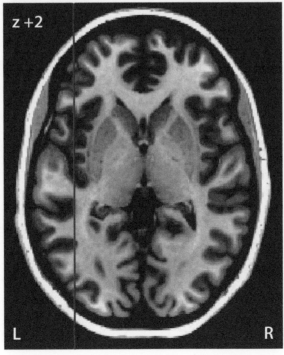

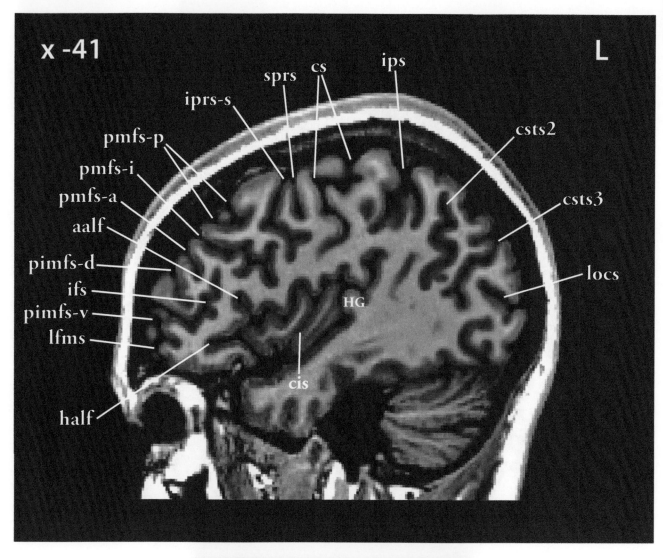

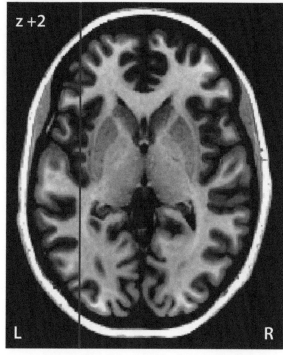

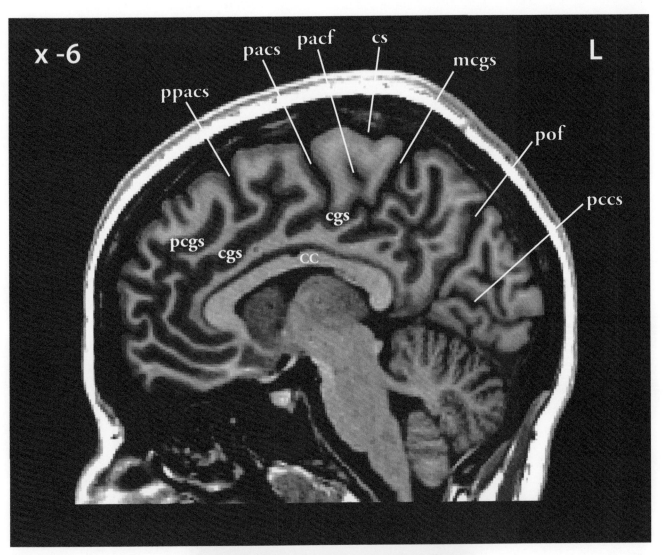

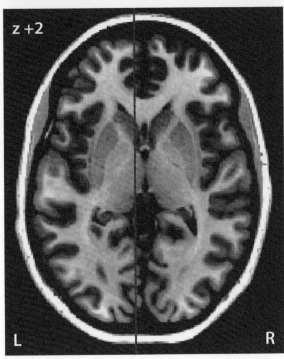

Cytoarchitecture

Cytoarchitecture

The cerebral cortex is an enormous sheet of gray matter consisting of neurons and glial cells. It is not a homogeneous structure and can be divided into many areas distinguished on the basis of their different cellular architecture (cytoarchitecture). These different areas are the functional units of the cerebral cortex in the sense that they receive inputs from specific cortical areas and subcortical structures, compute information, and send their output to the areas with which they are linked. They are relevant to understanding the neural basis of language, because all cognitive processing is nothing more than the complex functional interactions between networks of cortical and subcortical areas. Thus, defining the cortical areas (through cytoarchitecture) that participate in language processing is of the essence. The present section of the atlas provides photomicrographs and a brief description of the key cytoarchitectonic features of the various cortical areas that form the core language system.

The subdivision of the cerebral cortex into areas on the basis of their cellular organization started in the latter half of the 19th century when it became possible to harden the brain, section it into very thin slices that are within the range of tens of microns, and stain its cellular elements. Staining of the axons and dendrites of the neurons gave rise to myeloarchitectonics, while staining of the cell bodies of the neurons and the glial cells in the cortex gave rise to cytoarchitectonics. Meynert (1867, 1885), a pioneer in this type of investigation, was able to demonstrate cellular differences between the rhinencephalic region and the neocortex and to describe various layers of neurons in the cortex of the calcarine sulcus. These early studies were followed by several others in the latter part of the 19th century (e.g., Betz, 1874; Lewis and Clarke, 1878). The first complete cytoarchitectonic map of the human cerebral cortex appeared at the beginning of the 20th century (Campbell, 1905), followed by Brodmann's map (1908, 1909), which eventually became the most famous map of the human cerebral cortex. Brodmann used numbers to refer to the different cytoarchitectonic areas. Later, Economo and Koskinas (1925) published

an outstanding atlas of the cytoarchitecture of the human cerebral cortex in which they used letters to label the various cortical areas: F for frontal areas, P for parietal, T for temporal, and O for occipital areas. Thus, FA, FB, FC, FD, etc, refer to different frontal cytoarchitectonic areas, while PA, PB, etc, refer to parietal areas.

In the 1950s, interest in cytoarchitecture declined, but it was revived in the 1970s with the development of methods which permitted the tracing of the axonal connections of the different parts of the cerebral cortex in the monkey. The revival of interest in cytoarchitecture was largely due to the observation that the various cytoarchitectonic areas had distinct cortical and subcortical connections (e.g., Pandya and Sanides, 1973). The latter half of the 20th century was the golden era of the anatomical investigation of the connections of the monkey brain because of the development of precise methods for examining such connections (e.g., Fink and Heimer, 1967; Cowan et al., 1972; Kuypers and Huisman, 1984).

With the emergence of modern functional neuroimaging in the 1980s and the demonstration of distinct foci of functional activity in the human brain, the map of Brodmann (1909) was used to describe the different activation foci in the human cerebral cortex because its nomenclature for the various cytoarchitectonic areas had already been incorporated into the Talairach and Tournoux (1988) stereotaxic space that was adopted by the functional neuroimaging community. The Talairach and Tournoux (1988) proportional stereotaxic space gradually evolved into the Montreal Neurological Institute (MNI) standard stereotaxic space (Collins et al., 1994; for review, see Collins, 2012). Several modern architectonic investigations of the human cerebral cortex have attempted to provide a description of the variability in the location of cortical areas in the Talairach and Tournoux (1988) and the MNI stereotaxic space. These probabilistic cytoarchitectonic studies were pioneered at the University of Juelich, Germany, by Zilles and Amunts (e.g., Amunts et al., 1999; Morosan et al., 2001; Rademacher et al., 2001;

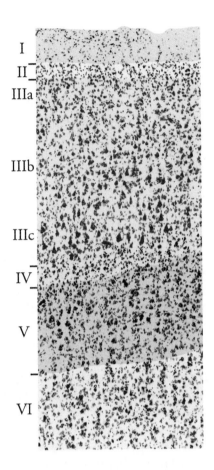

I

II

IIIa

IIIb

IIIc

IV

V

VI

FIGURE 17 Illustration of the six layers of a typical cortical area based on the staining of the cell bodies. Variations in these layers are the basis for cytoarchitectonic analysis and definition of the various cortical areas.

Caspers et al., 2006).

In the phylogenetically newer cortex, the so-called neocortex or isocortex, which comprises most of the primate cerebral cortex, six layers of neurons can be identified (Fig. 17). It should be realized that this is a convention and that these six layers can be further divided into sub-layers. Cytoarchitectonic analysis is based on the selective staining with various dyes (e.g., thionine, cresyl violet) of the bodies of the neurons and glial cells. These stains, which are collectively known as Nissl stains, permit an examination of the cellular organization of a particular part of the cerebral cortex. Counting from the surface of the cortex towards the white matter, the six layers of neurons are identified with the Latin numerals I to VI (Fig. 17). Layer I (molecular layer) has very few cell bodies, consisting mostly of neuroglial cells and non-pyramidal neurons. Layer II (external granular layer) has a dense population of mostly very small pyramidal neurons and, therefore, the name 'external granular layer' results from the overall granular appearance of this layer. Layer III (external pyramidal layer) is a broad layer of pyramidal neurons. There is a gradual increase in the size of the neurons of layer III, being smaller on top (just below layer II) and quite large in the lower part of this layer, just above layer IV. This layer is sometimes subdivided into three parts, sublayer IIIa referring to the smallest neurons just below layer II, sublayer IIIb to the middle part of the layer, and sublayer IIIc referring to the largest neurons just above layer IV. Layer IV (internal granular layer) is a narrow densely packed population of very small neurons, most of which are granular. Layer V (internal pyramidal layer) consists primarily of pyramidal neurons that are rather loosely dispersed. The neurons in the lower part of layer V are often less densely packed than those of the upper part, thus giving the appearance of an upper, Va, and a lower, Vb, sublayer. Layer

VI (multiform or polymorph layer) consists of more densely packed neurons (relative to those of sublayer Vb) that are a modified form of pyramidal cell.

Although the details of the physiological significance of the various layers are not clear, certain facts have emerged. The dendrites of the small granular (stellate) neurons that comprise layer IV are the targets of axons from subcortical nuclei, such as the thalamus and other cortical areas. The short axons of the stellate neurons do not leave the cortical area but rather distribute information to neurons in other layers. Thus, layer IV is primarily a locus within a cortical area for receiving information. The basal dendrites of the large neurons in the deeper part of layer III extend into layer IV and are, therefore, also the targets of inputs to this layer. The axons of layer III pyramidal neurons enter the white matter and course to other cortical areas within the same hemisphere, forming the cortico-cortical association pathways that will be described in the next section of the atlas (Connectivity of the Core Language Areas). These neurons are also the origin of the axons that course to the contralateral hemisphere to terminate in the corresponding cortical area and thus form the commissural fiber systems, i.e. the corpus callosum and the anterior commissure. Other layers such as layer II and layer V may contribute some axons to the callosal fiber systems, but the extent and details of such a contribution are not known. The axons of layer V neurons project to various subcortical areas and in this manner form the cortico-subcortical white matter pathways that target the basal ganglia, the nonspecific thalamic nuclei, and various other nuclei in the diencephalon and the brainstem. The axons arising from the motor cortical areas and reaching the spinal cord, forming the cortico-spinal pathways, also originate from layer V neurons. We do not know whether the axons targeting various subcortical regions originate from the same population of neurons or different populations within the same area. The axons of the Layer VI neurons course in the white matter to terminate in the main thalamic nucleus that sends input to that particular cortical area.

Many of the variations in the cellular structure of areas are relatively subtle, but there are examples of extreme differences. One extreme variation is observed in the cellular architecture of the fir (so-called primary) cortical areas that receive pa ticular types of sensory input, visual, auditory, somaesthetic, from the thalamus. Layer IV, th layer of small granular neurons that is the locus o the thalamic input, increases in width and layer II neurons that are normally significantly larger tha those of layer IV decrease in overall size. Thus, th boundaries between the small-celled layers II and IV and layer III become blurred, giving an overal granular appearance to this type of cortex. Unde low-power microscopy, these cortical areas have a overall dusty appearance that led Economo and Koskinas (1925) to introduce the term "koniocortex" to describe this, overall, "dusty" looking cortex (from the Greek words for "dust" κόνις and "dusty" κόνιος). Thus, the koniocortical primary visual, auditory, and somaesthetic sensory areas can be readily identified by their overall dusty appearance.

At the other extreme in cellular architecture are the motor areas found on the precentral gyrus and the adjacent posterior parts of the superior and middle frontal gyri. In these motor areas, the small granular (stellate) neurons interposed between layers III and V have disappeared. The cellular landscape in these areas is dominated by large pyramidal neurons in layer V with axons that reach subcortical targets at long distances. The lack of the typical granular layer IV leads to the description of these areas as "agranular". Thus, the cytoarchitectonic picture reflects to a certain extent the nature of the inputs and outputs of a particular cortical area.

Distinct cytoarchitectonic areas are found in the ventrolateral frontal, inferior parietal, superolateral temporal, and insular regions that constitute the core language regions of the cortex. The present section of the atlas provides photomicrographs of these areas and short, focused descriptions of their essential cytoarchitectonic features. The photomicrographs were taken at 10x magnification. The location of the various cytoarchitectonic areas found in the peri-Sylvian region is indicated on the 3-dimensional reconstruction of the average human brain in the Montreal Neurological Institute space (Fig. 18).

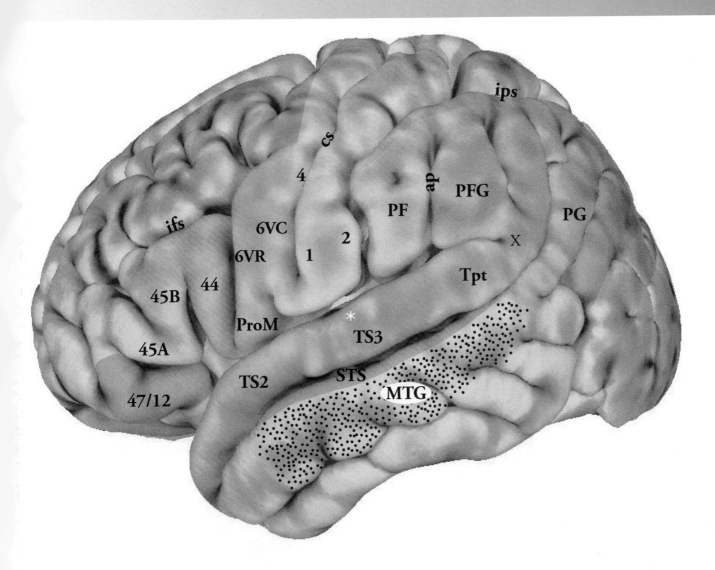

FIGURE 18 The location of the various cytoarchitectonic areas in the peri-Sylvian region is indicated on the 3-dimensional reconstruction of the average human brain in the Montreal Neurological Institute space. Abbreviations: ap, ascending posterior ramus of the lateral fissure; cs, central sulcus; ifs, inferior frontal sulcus; ips, intraparietal sulcus; x, parieto-temporal isthmus.

THE INFERIOR FRONTAL GYRUS: AREAS 44, 45 AND 47/12

The posterior part of the inferior frontal gyrus is traditionally referred to as Broca's Region. Cytoarchitectonically, this region comprises two distinct areas: area 44 and area 45 (Brodmann, 1909) (Fig. 19). Area 44 lies on the pars opercularis and area 45 on the pars triangularis of the inferior frontal gyrus (e.g., Amunts et al., 1999, 2010; Petrides and Pandya, 1994, 2002). On the pars orbitalis of the inferior frontal gyrus lies area 47/12. Although area 47/12 has not been traditionally treated as a core language area, functional neuroimaging has suggested that it may play a role in controlled access to stored conceptual representations (Badre and Wagner, 2007) and semantic unification (Zhu et al., 2012).

AREA 44 (AREA FCBm)

The cortex that lies on the pars opercularis of the inferior frontal gyrus has been referred to as area 44 by Brodmann (Fig. 19) and as area FCBm by Economo and Koskinas (Fig. 20). This cortical area is in continuity, posteriorly, with the most rostral

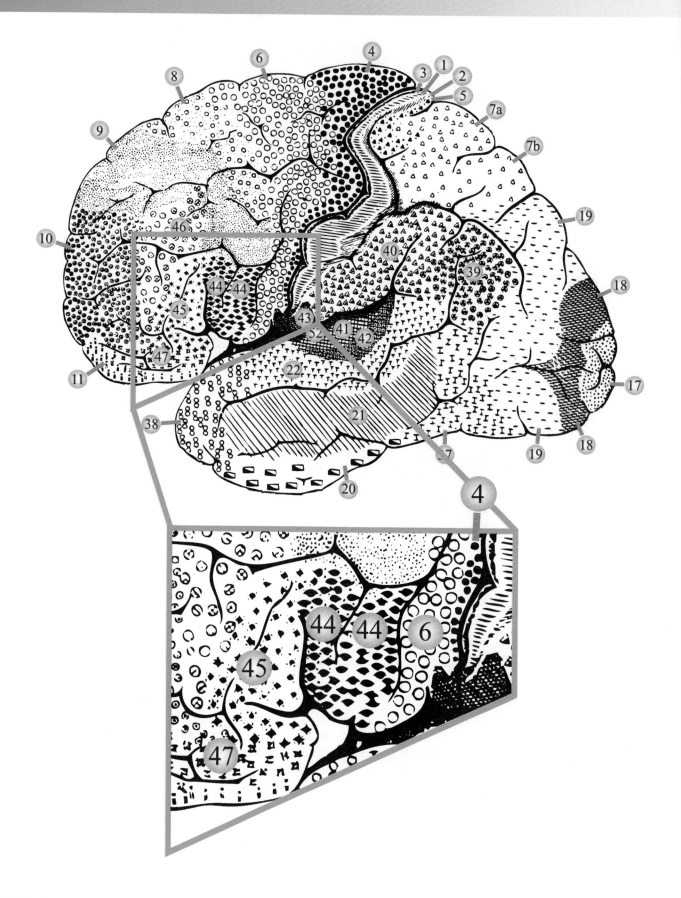

FIGURE 19 Brodmann's cytoarchitectonic map with the region around the inferior frontal gyrus expanded (inset) to highlight the location of architectonic areas 44 and 45 that comprise Broca's region and the adjoining areas.

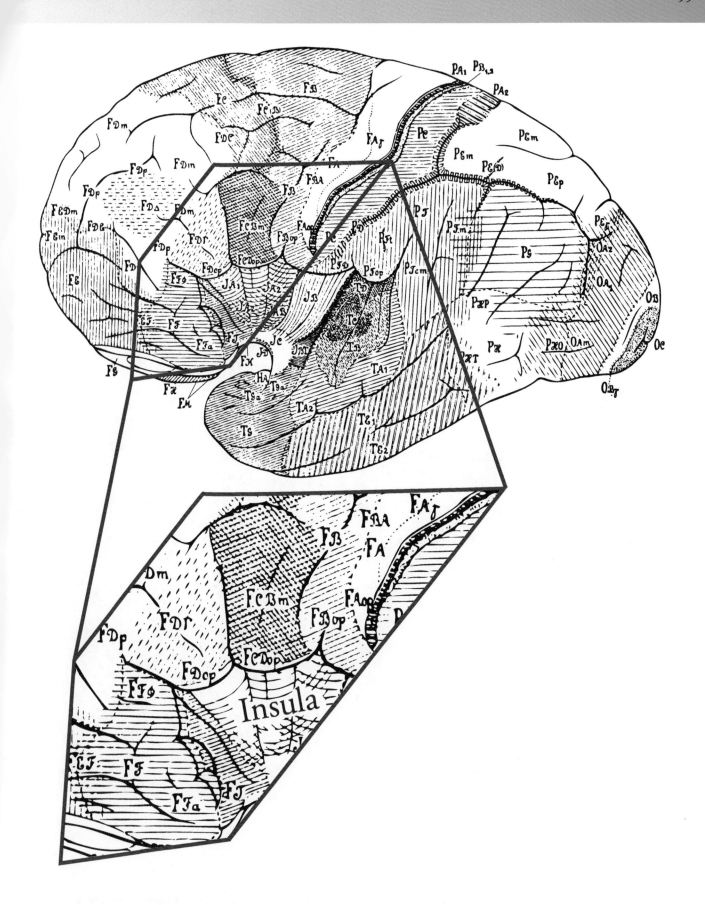

FIGURE 20 Cytoarchitectonic map of Economo and Koskinas (1925) with the region around the inferior frontal gyrus expanded (inset) to highlight the architectonic areas that comprise Broca's region and the adjoining areas.

Area 44

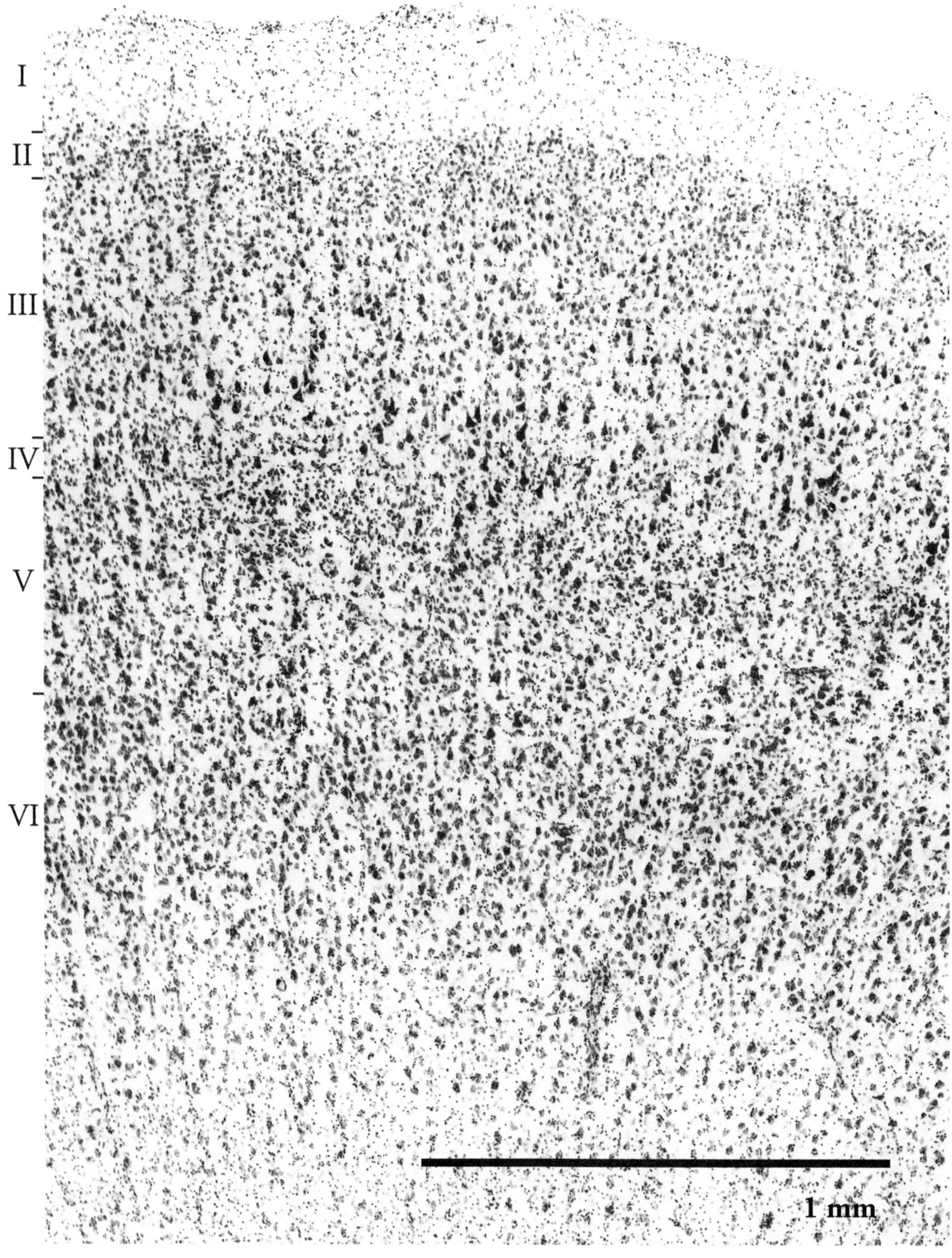

I

II

III

IV

V

VI

1 mm

Area 44 (layer IV highlighted)

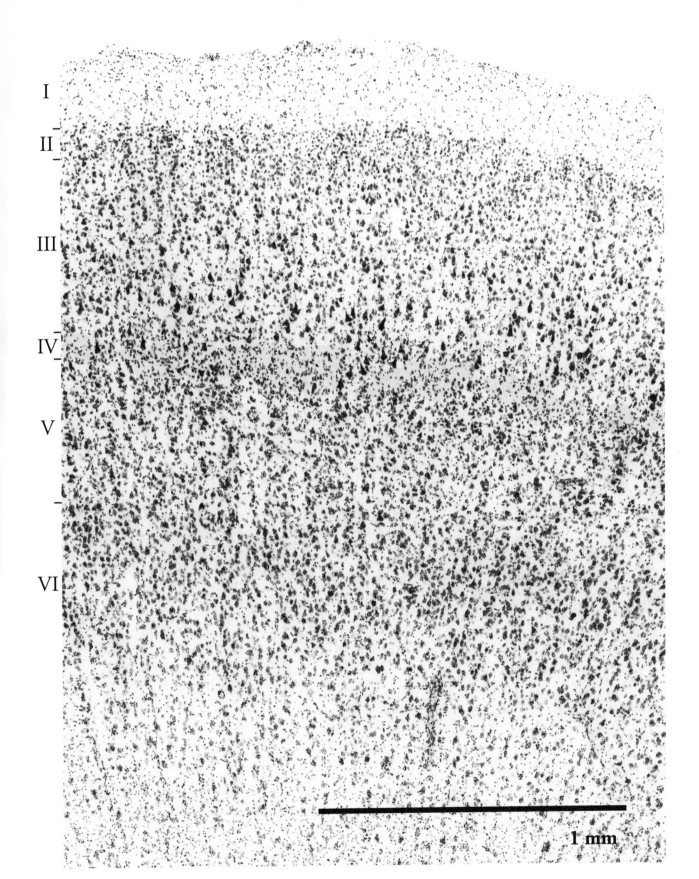

I

II

III

IV

V

VI

1 mm

Area 44 Op

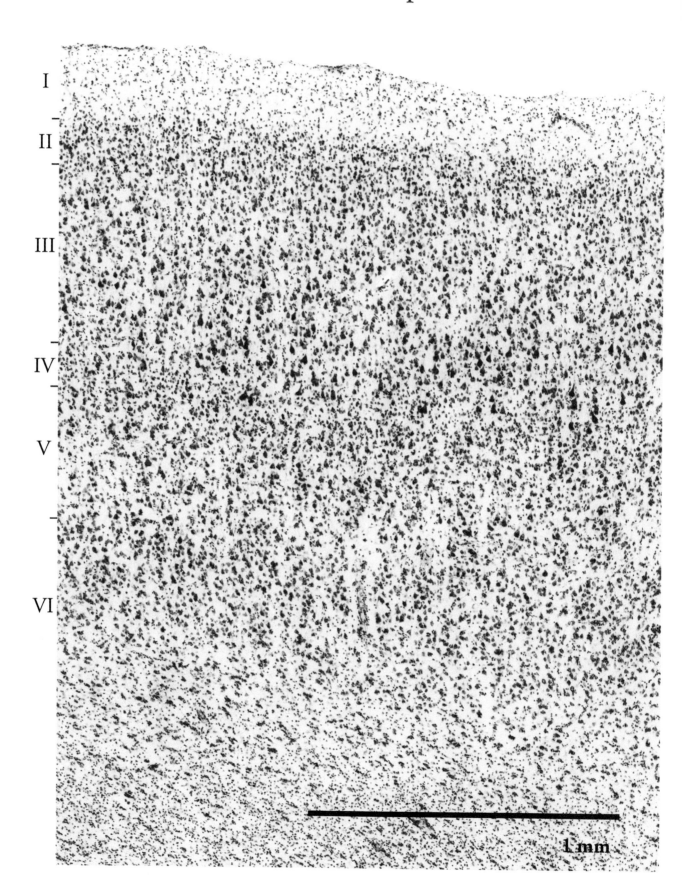

I

II

III

IV

V

VI

1 mm

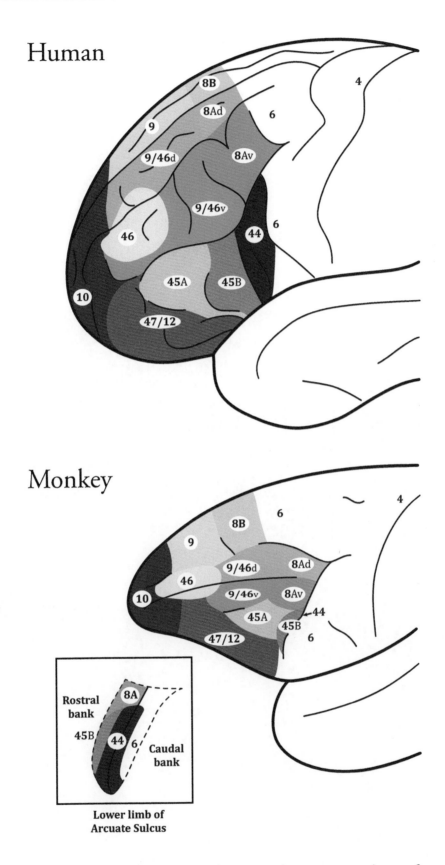

FIGURE 21 Comparative cytoarchitectonic map of the human and macaque monkey prefrontal cortex by Petrides and Pandya (1994, 2002). A part of the lower limb of the arcuate sulcus of the macaque monkey has been opened to show area 44 that lies in the fundus of this sulcus (see inset).

Area 45A

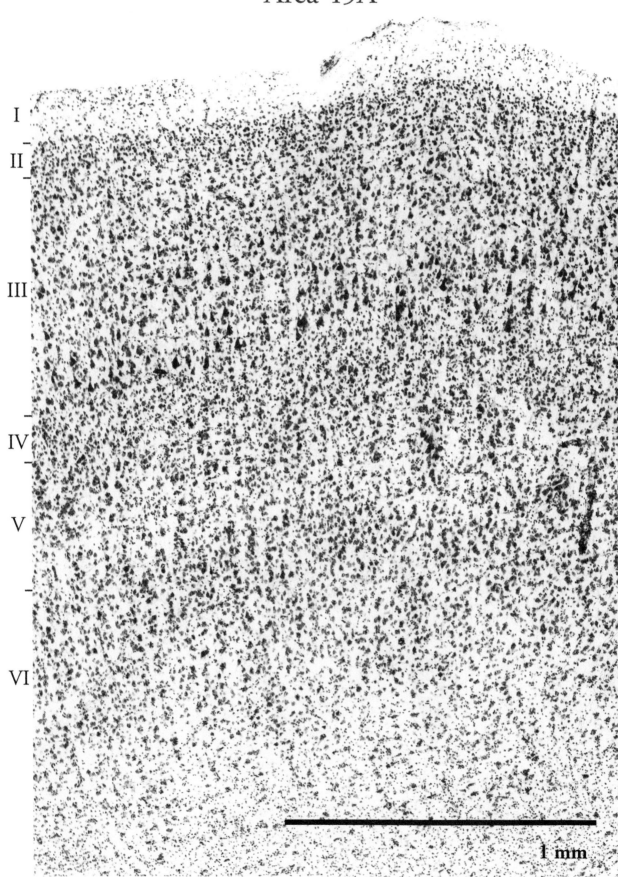

I

II

III

IV

V

VI

1 mm

Area 45A

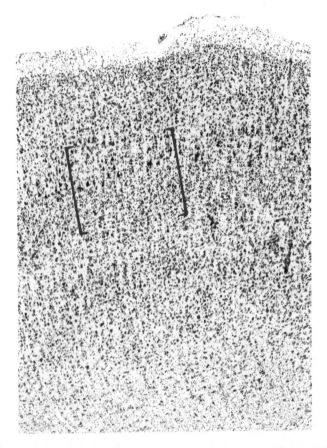

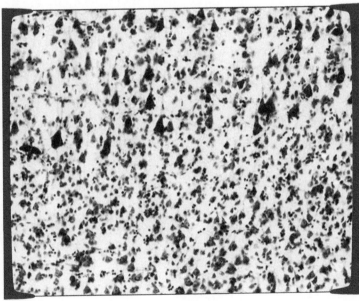

FIGURE 22 Photomicrograph of cytoarchitectonic area 45A with a part of layers III, IV, and V highlighted by the red brackets and expanded to illustrate the very large and deeply stained neurons in layer III, and the medium size neurons of layer V (just below layer IV). The combination of these features distinguishes this area from all other frontal granular areas.

Area 45A

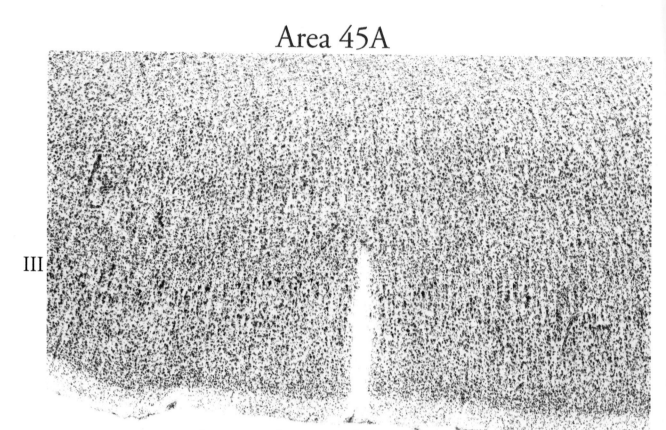

III

horizontal ramus of the lateral fissure

III

Area 47/12

Area 45A   Area 47/12

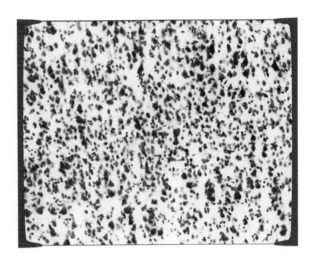

FIGURE 23 (LEFT) Panoramic view of the cortex across a section of the horizontal ramus of the lateral fissure to show area 45A on the upper bank and area 47/12 on the lower bank of the sulcus. In this photomicrograph, III marks the deepest part of layer III where the fundamental difference between these two areas is observed: presence of clusters of unusually large neurons in area 45A, but lack of these neurons in layer 47/12.

FIGURE 24 (ABOVE) Comparison of the deepest part of layer III, layer IV, and upper part of layer V of areas 45A and 47/12. Note the presence of the intensely stained and very large pyramidal neurons in the deepest part of layer III in area 45A and their absence in area 47/12. Note also the well-developed layer IV and the small to medium size neurons in layer V in both areas.

---

part of the precentral area 6, a typical agranular premotor area lacking layer IV. In area 44, layer IV is a thin and interrupted layer of small granular neurons. The interruptions of this layer are created by the invasion of the adjacent pyramidal neurons of layers III and V. In the photomicrograph of area 44 on p. 97, the thin and interrupted layer IV has been highlighted in yellow. The deepest part of layer III contains many very large pyramidal neurons. Layer V contains medium to large pyramidal neurons that are, overall, larger than those of Layer VI.

The emerging but not fully developed layer IV is the reason why area 44 is described as dysgranular and it is the fundamental architectonic characteristic that distinguishes it from the caudally adjacent agranular premotor area 6 and the rostrally adjacent prefrontal cortical area 45 which has a well

developed layer IV (Petrides and Pandya, 2002). One can therefore conclude that area 44 is a transitional cortical area between the orofacial part of premotor area 6 and the ventral prefrontal cortex.

At the base of the pars opercularis, a variation of area 44 is encountered and we refer to this variation as area 44 Op (see photomicrograph of this opercular area 44 on p. 98). Economo and Koskinas (1925) also recognized this opercular variation, as area FCDop (Fig. 20).

AREA 45 (AREA FDΓ)

Area 45, which occupies the pars triangularis of the inferior frontal gyrus (Figs. 18, 19, and 21), is typical prefrontal granular cortex: it has a well developed layer IV. This is a continuous layer of

Area 47/12

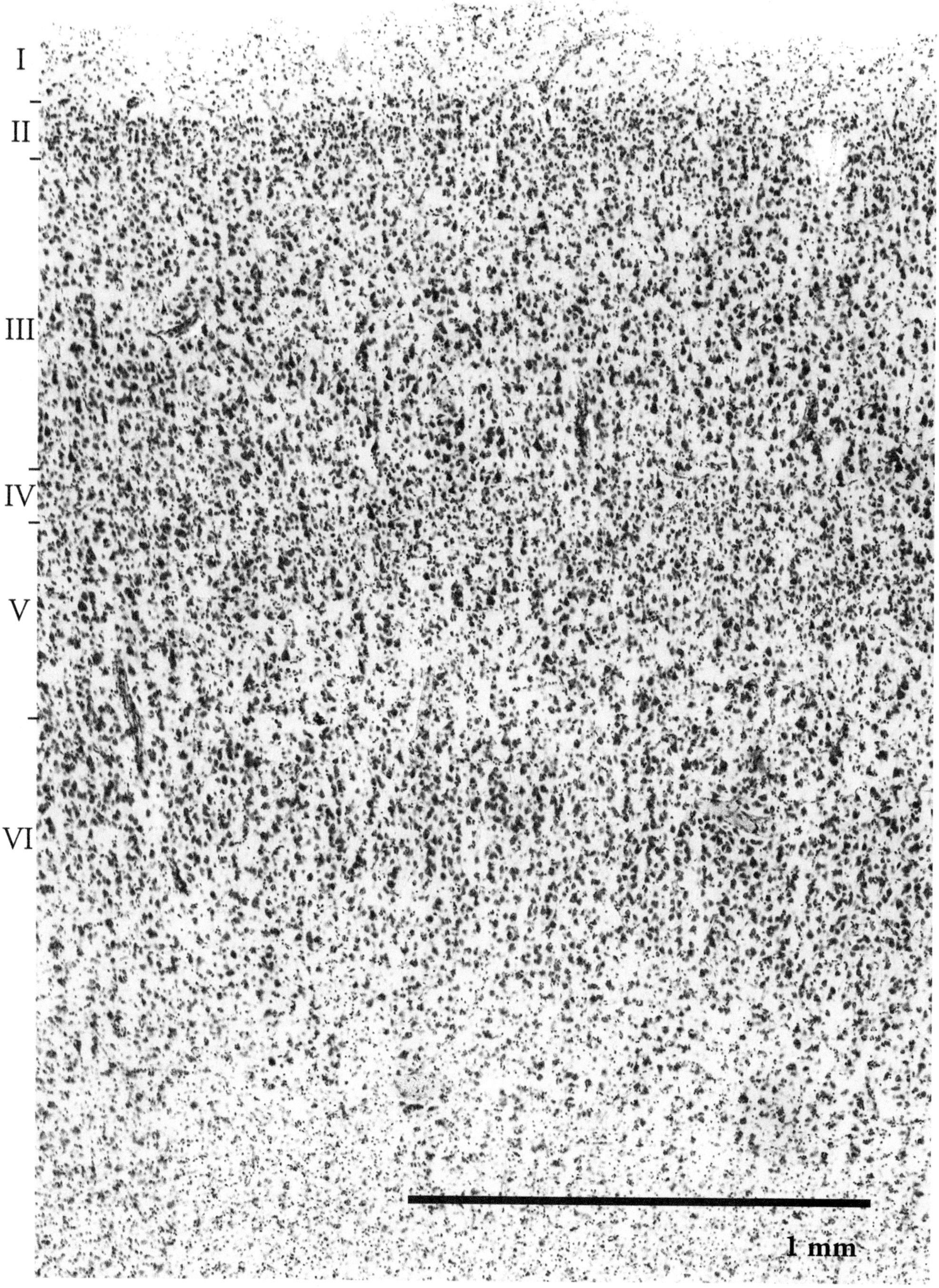

I

II

III

IV

V

VI

1 mm

Area 47/12

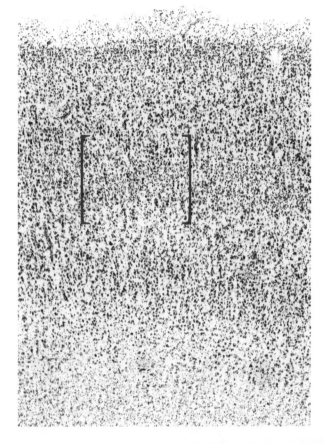

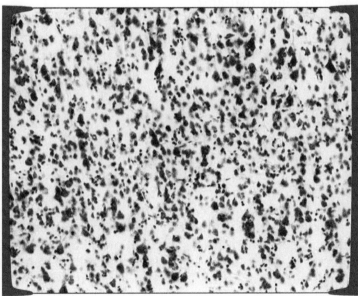

FIGURE 25 Photomicrograph of cytoarchitectonic area 47/12 with a part of layers III, IV, and V highlighted by the red brackets and expanded to illustrate the lack of very large neurons in layer III in this cortical area.

small stellate neurons that clearly separate the pyramidal neurons of the deepest part of layer III from those of layer V (Fig. 22). A key architectonic feature of area 45 is the presence of many intensely stained and unusually large pyramidal neurons in the deepest part of layer III (Fig. 22). Economo and Koskinas (1925) refer to this area as FDΓ (Fig. 20), the Greek capital letter gamma (Γ) being used to signify the presence of gigantic pyramidal neurons in layer III (Γίγας, for "giant" in Greek). Layer V has pyramidal neurons of moderate size. The unusually large neurons in the deep part of layer III distinguish this area from all surrounding prefrontal areas and the well developed layer IV distinguishes it from caudally adjacent area 44 (Petrides and Pandya, 2002).

Area 47/12

The cortex on the pars orbitalis of the inferior frontal gyrus has often been discussed in relation to its possible contribution to language processing. The cytoarchitectonic description of this cortical region has generated a lot of confusion. Inspection of the map of Brodmann (1909) shows that area 45 is replaced, ventrally, by area 47 (Fig. 19). The term "area 47" was used loosely by Brodmann to refer to a heterogeneous region that occupies the entire caudal extent of the orbital frontal surface. Brodmann (1909) commented on the heterogeneity of this region and pointed out that he did not differentiate its various parts. Economo and Koskinas (1925) referred to this broad caudal orbitofrontal region as area FF (Fig. 20) and also remarked on its cytoarchitectonic heterogeneity. The architectonic landscape in this large caudal orbital frontal region (referred to as area 47 by Brodmann and as area FF by Economo and Koskinas) ranges from an agranular part close to the caudal gyrus rectus (i.e. a part of area 47 lacking layer IV) to a dysgranular part more laterally and, finally, to a fully granular part on the most lateral part of the orbital surface where it continues into the pars orbitalis of the inferior frontal gyrus. Economo and Koskinas (1925) indicate a transition of area FDΓ to a distinct granular area named FFφ (Fig. 20). Petrides and Pandya (1994, 2002) also observed a distinct transition from area 45 (area FDΓ) to another granular area that lies below the horizontal ramus of the lateral fissure (Fig. 23). They labelled this area 47/12 (Fig. 21) because it corresponds to only the most lateral and granular part of Brodmann area 47 and, in the monkey, corresponds to the ventrolateral frontal area labelled as area 12 by Walker (1940). The remaining dysgranular and agranular parts of Brodmann area 47 that spread into the caudal orbitofrontal region were labelled as area 13 by Petrides and Pandya (1994, 2002) to be consistent with the homologous region in the macaque monkey.

As can be seen in figures 23 and 24, both areas 45A and 47/12 are granular frontal cortex with a well-developed layer IV. The distinction between area 45 and area 47/12 can easily be made because area 47/12 lacks the unusually large neurons that are found in the deepest part of layer III in area 45 (Figs. 23 and 24), permitting a reliable placing of the border between these two areas. A photomicrograph of area 47/12 is provided in figure 25.

The Precentral Gyrus: Motor Areas 4, 6VC, 6VR

Broca's region on the inferior frontal gyrus lies just anterior to the motor cortex of the ventral precentral gyrus where the orofacial musculature is represented (Penfield and Boldrey, 1937; Woolsey et al., 1979). Brodmann made a distinction between two areas on the precentral gyrus, areas 4 and 6 (Fig. 26). Area 4 occupies most of the gyrus dorsally, close to the midline, but recedes and stays close to the central sulcus more ventrally. In other words, in the ventral part of the precentral gyrus where the orofacial musculature is represented, area 4 is largely confined to the anterior bank of the central sulcus and occupies only a small part of the crown of the gyrus (Fig. 26). Although Brodmann (1909) identified only two cytoarchitectonic areas (4 and 6) on the ventral precentral gyrus, Economo and Koskinas (1925) identified three areas: area FAγ which corresponds to Brodmann area 4 and areas FA and FB which correspond to the caudal and anterior parts of Brodmann area 6 (Fig. 27). Modern monkey architectonic studies also identify three areas on the ventrolateral precentral region labeled as area F1 (close to the central sulcus),

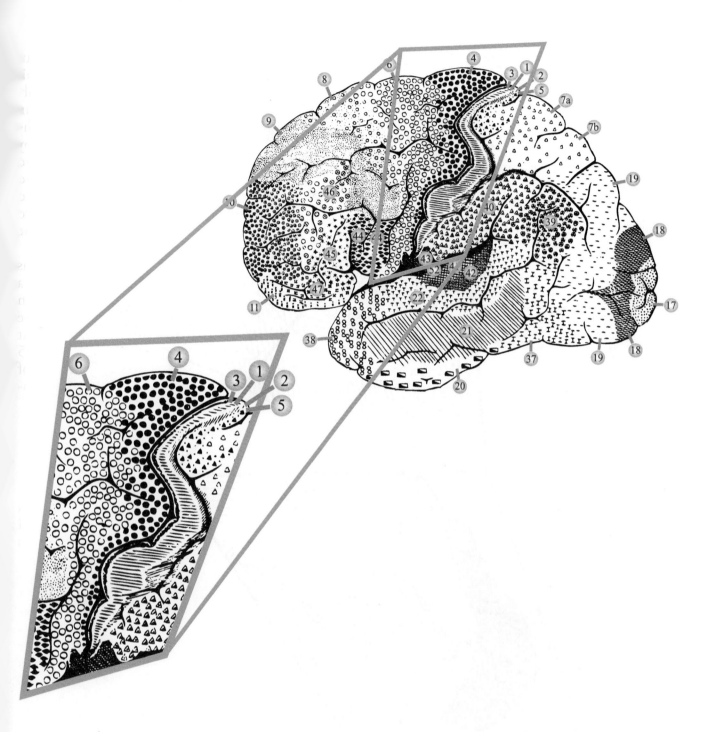

FIGURE 26 Brodmann's cytoarchitectonic map with the region around the central sulcus expanded (inset) to highlight the location of architectonic areas 4 and 6 on the precentral gyrus and areas 3, 1, and 2 on the postcentral gyrus.

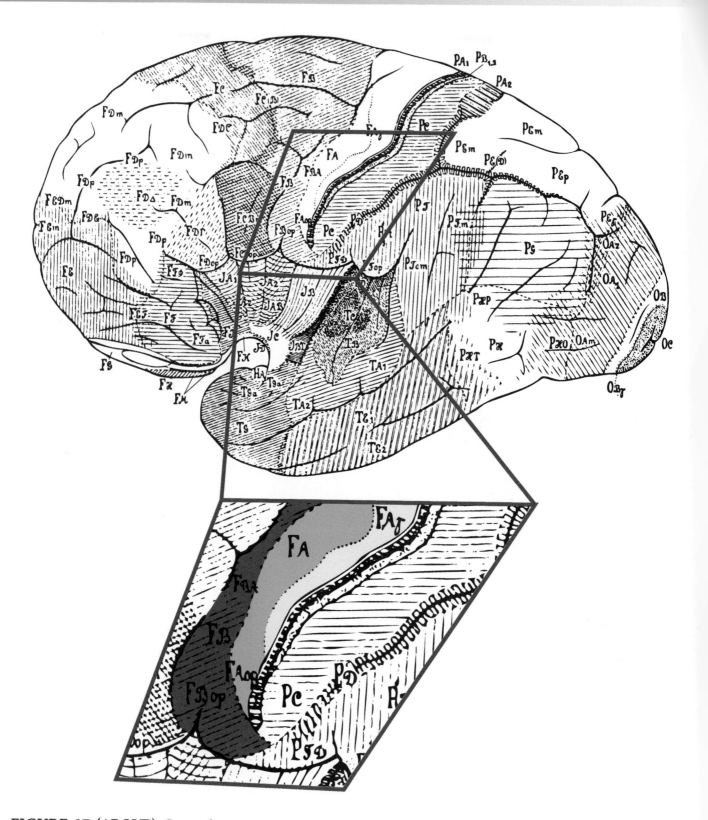

FIGURE 27 (ABOVE) Cytoarchitectonic map of Economo and Koskinas (1925) with the region around the ventral part of the precentral gyrus expanded (inset) to highlight the architectonic areas that comprise this region.

FIGURE 28 (RIGHT) Photomicrographs of primary motor area 4 and primary somatosensory area 3b taken from the same point in the ventral part of the central sulcus. Area 4 occupies the anterior bank and area 3b the posterior bank of the central sulcus. Note the striking difference in the thickness of these two areas.

Area 3b

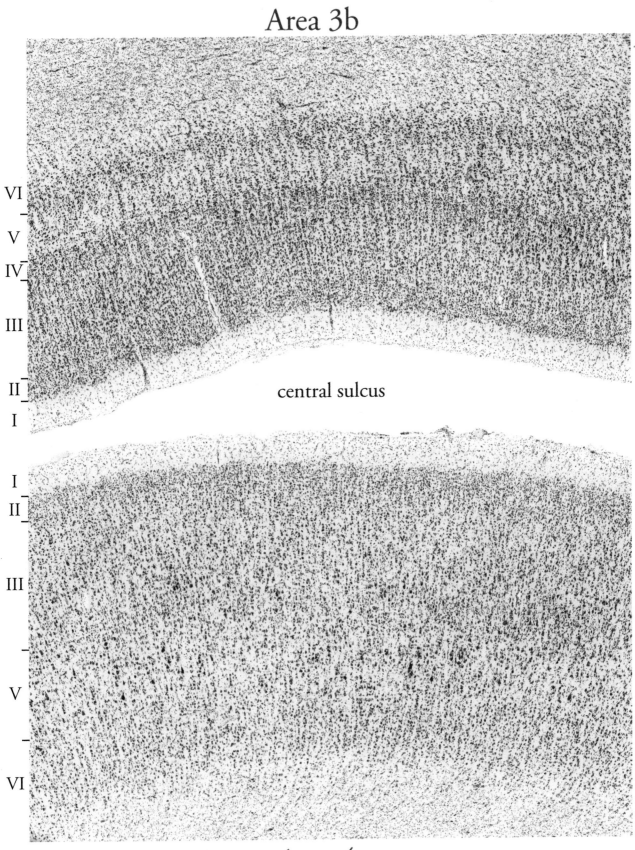

central sulcus

Area 4

Area 4 Orofacial

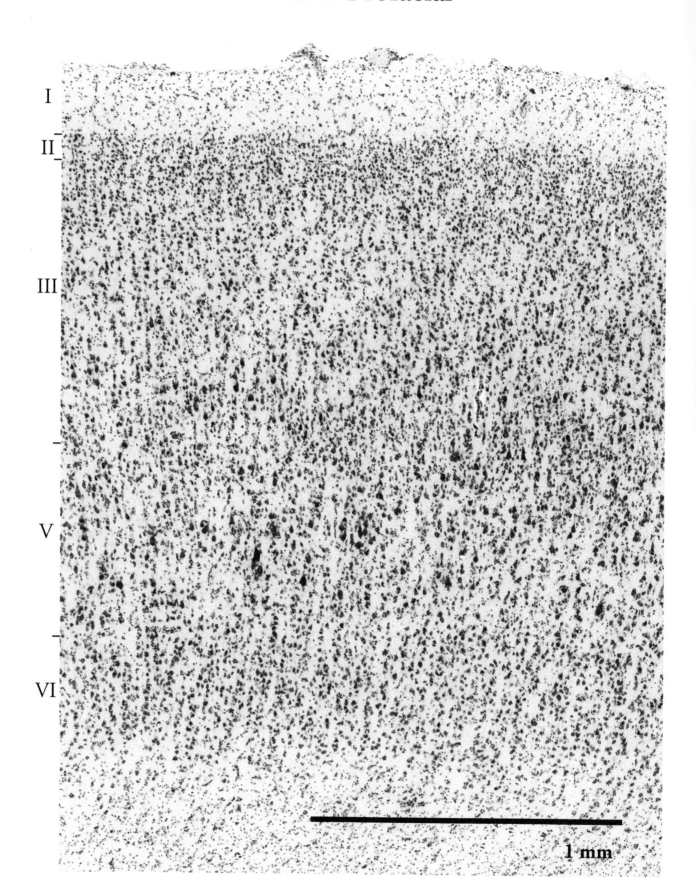

I

II

III

V

VI

1 mm

Area 3b

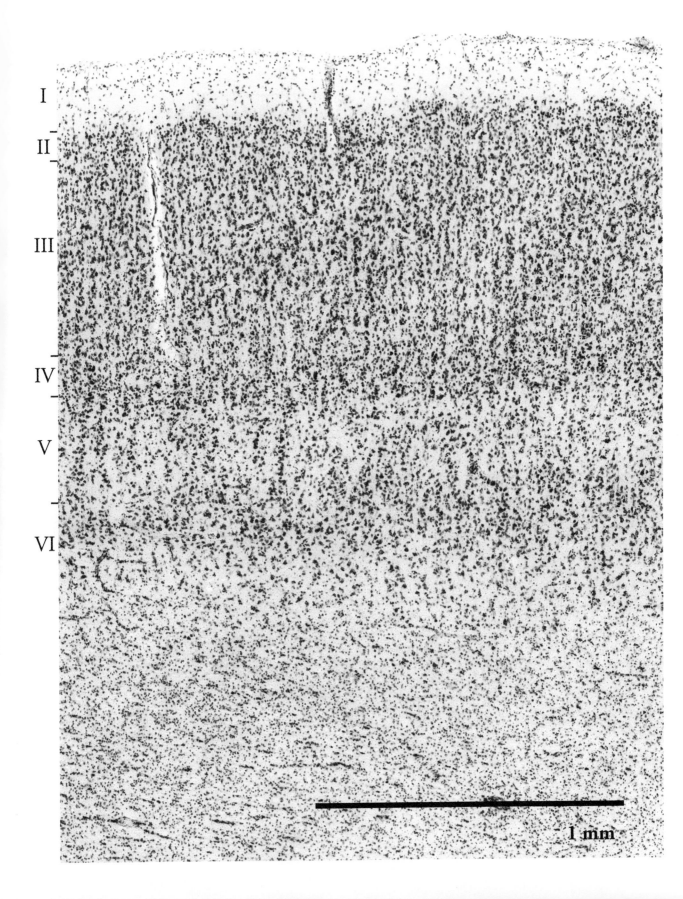

I

II

III

IV

V

VI

1 mm

Area 4 Orofacial, Hand, Foot

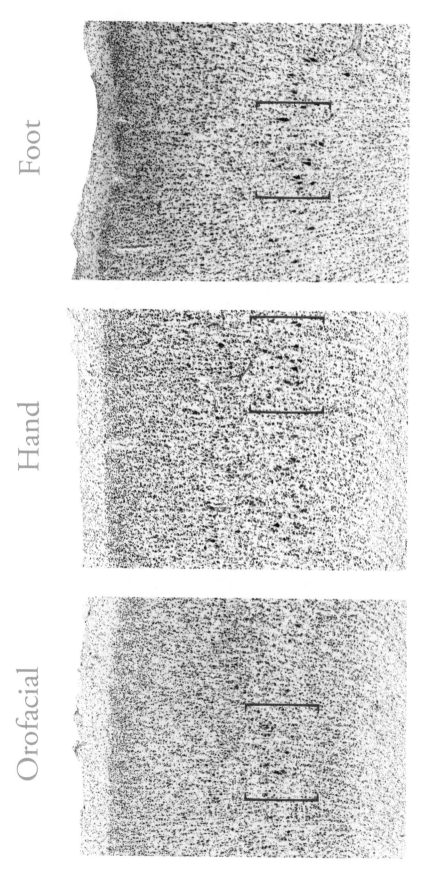

Foot

Hand

Orofacial

Area 4 Orofacial, Hand, Foot

Orofacial   Hand   Foot

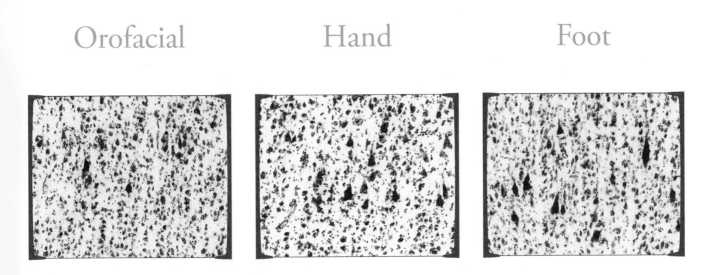

FIGURE 29 (LEFT AND ABOVE) Photomicrographs of area 4 taken from the medial surface of the brain (leg and foot representation), the middle section of the precentral gyrus (arm and hand representation) and the most ventral part of the precentral gyrus (orofacial representation). The red brackets outline a part of the cortex where the Betz neurons are found. Note that these Betz neurons are largest in the region of the leg and foot, smaller in the region of the arm and hand, and even smaller in the orofacial region.

---

followed by area F4, and then area F5 at the most anterior part of the ventral precentral region (e.g., Matelli et al., 1985; Rizzolatti and Luppino, 2001).

Area 6 of Brodmann covers the anterior part of the precentral gyrus (in front of area 4) and extends anterior to the superior precentral sulcus to cover the caudal parts of the superior and middle frontal gyri as well as the caudal part of the medial surface of the superior frontal gyrus (Fig. 26). Modern studies in the macaque monkey have subdivided Brodmann's area 6 into six distinct motor areas (e.g., Rizzolatti and Luppino, 2001). Area 4 and all subdivisions of area 6 are "agranular" cortical areas. The term "agranular" refers to the absence of layer IV, namely the layer of small granular (stellate) neurons that are interposed between the pyramidal neurons of layers III and V. In the space normally occupied by layer IV, small pyramidal neurons can be observed.

AREA 4 (FAγ)

Area 4 of Brodmann or area FAγ of Economo and Koskinas, the primary motor cortical area, is found on the posterior part of the precentral gyrus close to the central sulcus (Figs. 26 and 27). The anterior bank of the central sulcus is always occupied by this motor cortical area (Fig. 28). Close to the midline, where the musculature of the trunk is represented, this area extends anterior to the central sulcus and covers most of the precentral gyrus (Figs. 26 and 27). Ventrally, where the orofacial musculature is represented, area 4 is virtually confined to the anterior bank of the central sulcus. The most striking characteristic of the agranular area 4 is the presence of extremely large neurons, known as the Betz cells, in the deeper part of layer V (see photomicrograph of area 4 on pages 110 and 117, as well as Fig. 29). The Betz neurons are found either individually or in groups of just

Area 6VR

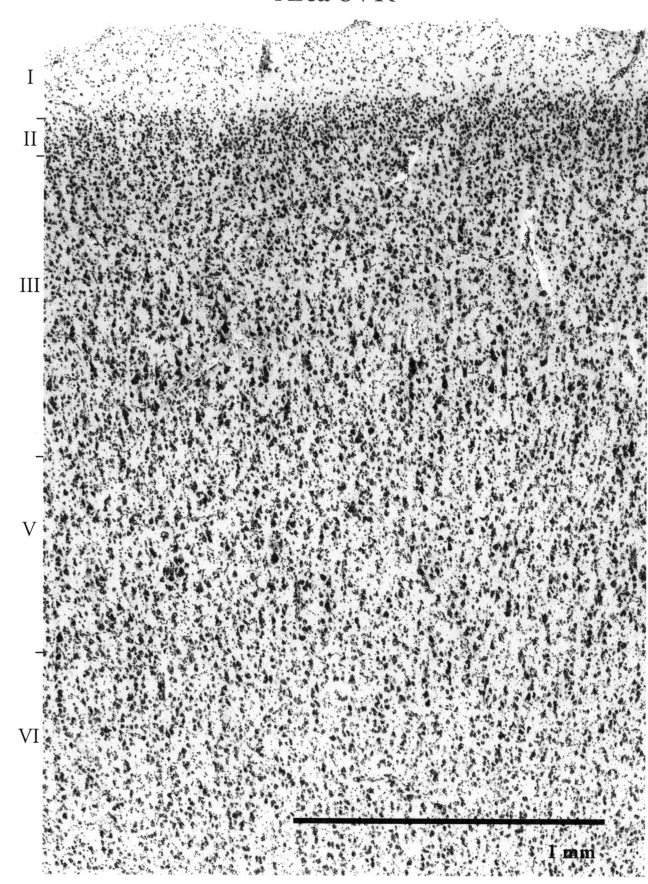

I

II

III

V

VI

1 mm

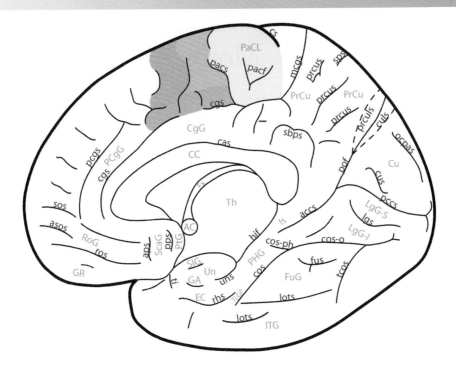

FIGURE 30 Schematic view of the medial surface of the hemisphere to highlight medial area 4 (yellow) on the paracentral lobule, the supplementary motor region (orange), and the adjoining cingulate motor region (blue). For abbreviations, see Abbreviations List.

a few neurons. The presence of these extremely large neurons led Economo and Koskinas (1925) to refer to this area as the giganto-pyramidal motor area FAγ. The Greek letter γ is used as a suffix to signify the presence of these giant (γίγας in Greek) neurons. The largest Betz neurons are observed in the dorsal and the medial part of area 4 where the trunk and legs are represented. These neurons decrease in size in the middle part of the precentral gyrus where the arm and hand are represented and their size decreases even further close to the lateral fissure where the orofacial part of the body is represented (Fig. 29).

A comparison of area 4 on the anterior bank of the central sulcus with the primary somatosensory area 3b on the posterior bank of this sulcus demonstrates a striking difference in the thickness of these areas (Fig. 28). In summary, the primary motor area 4 is a thick, agranular area that is uniquely characterized by the presence of the gigantic Betz neurons in layer V.

AREA 6VC (AREA FA) AND AREA 6VR (AREA FB)

Area FA of Economo and Koskinas (Fig. 27) covers a large part of the ventral precentral gyrus immediately anterior to area FAγ (area 4). Area FA and area FAγ are almost identical agranular motor areas, except that area FA lacks the gigantic Betz neurons in layer V. Area FA is the caudal part of the cortex that Brodmann labelled area 6 and its ventral component can, therefore, be called area 6VC (VC for ventrocaudal). Its dorsal part can be referred to as 6DC (DC for dorsocaudal).

Anterior to area 6VC lies the rostral part of ventral area 6, area 6VR, which corresponds to the ventral part of area FB of Economo and Koskinas (Fig. 27). The agranular premotor area 6VR continues into the posterior bank of the inferior precentral sulcus. Layers III and V contain large pyramidal neurons of about the same size, making it difficult to draw the boundary between these two layers (see photomicrograph of this area on p. 114). Under the relatively distinct line of large neurons that can be observed in the deepest part of layer III lie medium size pyramidal neurons, followed

Supplementary Motor Area (SMA)

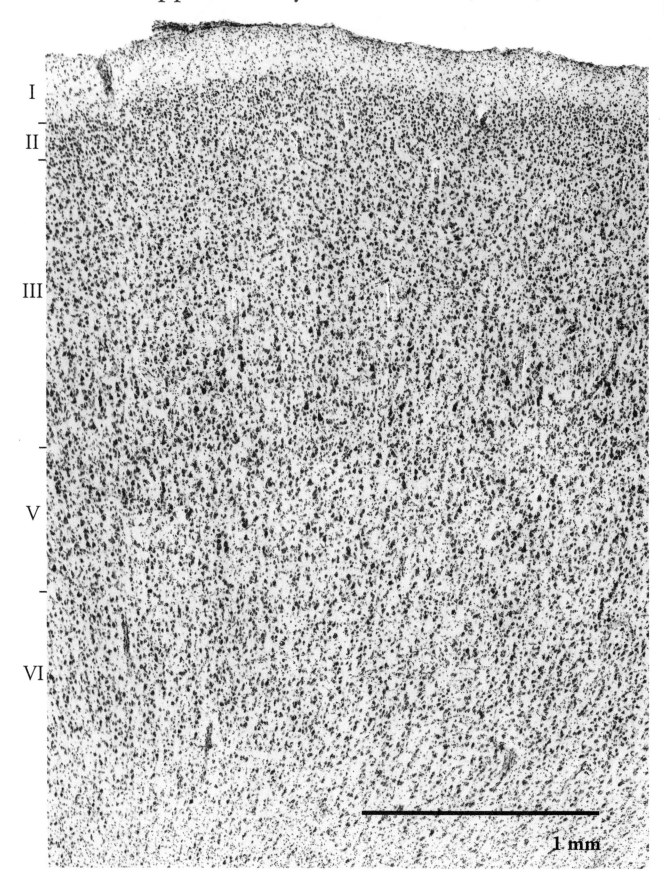

I

II

III

V

VI

1 mm

Area 4 (medial surface)

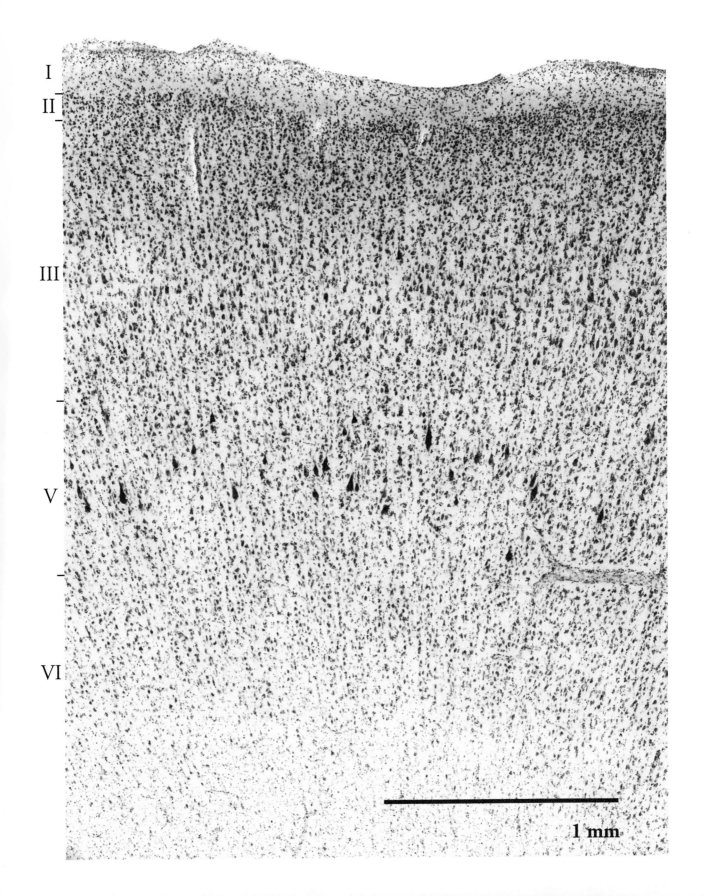

I

II

III

V

VI

1 mm

Area 2

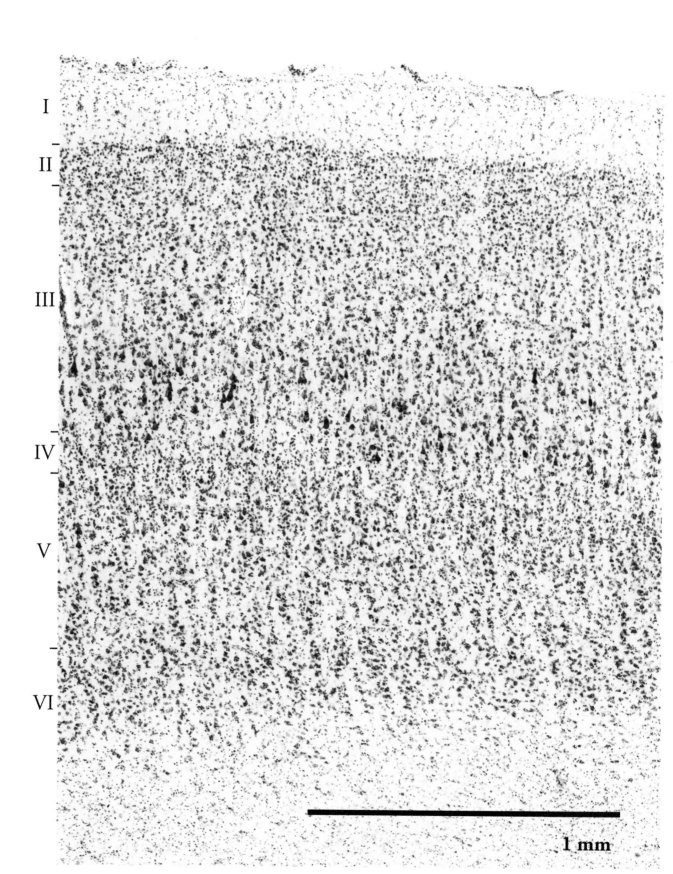

I

II

III

IV

V

VI

1 mm

Area 2

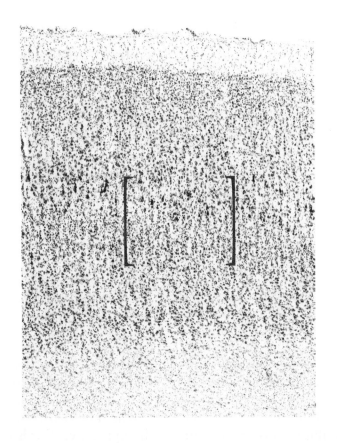

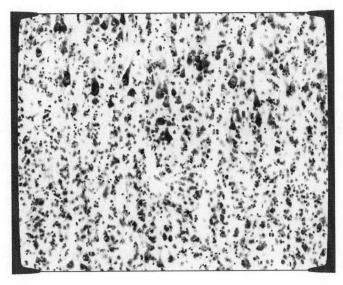

FIGURE 31 Photomicrograph of cytoarchitectonic area 2 with a part of layers III, IV, and V highlighted by the red brackets and expanded to illustrate the large pyramidal neurons in layer III and the medium size neurons of layer V (just below layer IV).

by the large neurons of layer V. These smaller medium size pyramidal neurons may at times give the illusion of some granularity, but examination under high-field microscopy shows that they are not granular (stellate) neurons but rather smaller pyramidal neurons.

SUPPLEMENTARY MOTOR REGION (MEDIAL AREA 6)

The central sulcus often continues for a short distance onto the medial surface of the hemisphere. Anterior and posterior to it lie the medial extensions of the precentral and postcentral gyri creating the morphological entity known as the paracentral lobule (yellow in Fig. 30). On the paracentral lobule lies the medial extension of area 4 (area FAγ) where the leg of the body is represented. Immediately anterior to the paracentral lobule lies the medial extension of area 6 (orange in Fig. 30). Electrical stimulation of this region during brain surgery demonstrated the existence of another representation of the body that led Penfield to name it "supplementary motor area" (Penfield and Welch, 1951). Stimulation of this region can lead to vocalization, speech interference, or speech arrest (Figs. 6-8) (Penfield and Roberts, 1959). Photomicrographs of the cytoarchitecture of the supplementary motor area and the medial part of area 4 that lies on the paracentral lobule are shown on pages 116 and 117 for comparison. Note the smaller scale bar in these two photomicrographs and the absence of the Betz neurons in the supplementary motor area.

THE POSTCENTRAL GYRUS: SOMATOSENSORY AREAS 3, 1 AND 2

The agranular and thick motor cortical area 4 that occupies the anterior bank of the central sulcus is followed by the markedly thin primar somatosensory cortical area 3b on the posterio bank of this sulcus (Fig. 28 and photomicrograph on pages 110 and 111). Area 3b is typical konio cortex in the sense that small granular neuron and small pyramidal neurons predominate givin this area an overall "dusty" appearance under lo power microscopic examination. The pyramida neurons of layer III are small and thus do not stanc out clearly against layers II and IV, as is the case ir other areas. Layer IV is very well developed. Laye V becomes significantly thin and is sparsely occu pied by small pyramidal neurons. In the fundu of the central sulcus where the transition between agranular motor cortex and hypergranular somatosensory cortex is occurring, one encounters a hybrid cortex, area 3a, which exhibits characteristics of both the primary motor agranular area 4 and the koniocortical area 3b: a well developed layer IV co-exists with Betz neurons in layer V. This hybrid area 3a continues for a while in the deeper part of the posterior bank of the central sulcus (approximately the lower third) before it is fully replaced by the koniocortical area 3b.

The koniocortical area 3b occupies most of the upper two thirds of the posterior bank of the central sulcus. On the crest of the postcentral gyrus, area 3b is replaced by a slightly less granular cortical area in which the neurons of the deeper part of layer III increase in size and layer V is not as sparse. This is area 1 of Brodmann (1909). The caudal part of the postcentral gyrus and both banks of the postcentral sulcus are occupied by area 2 (Fig. 26). This somatosensory area has a well developed granular layer IV, but unlike areas 3b and 1, has large pyramidal neurons in the deeper part of layer III and medium size neurons in layer V (see Fig. 31). Figure 32 provides a comparison of the three somatosensory areas that are found on the postcentral gyrus. Note the difference in thickness between the three areas and the relatively empty layer V in koniocortical area 3b.

FIGURE 32 Photomicrographs of the three cytoarchitectonic areas that occupy the postcentral gyrus: area 3b, area 1 and area 2. These three distinct areas constitute the somatosensory region of the postcentral gyrus. Scale bar, 1mm.

Areas 3b, 1, and 2

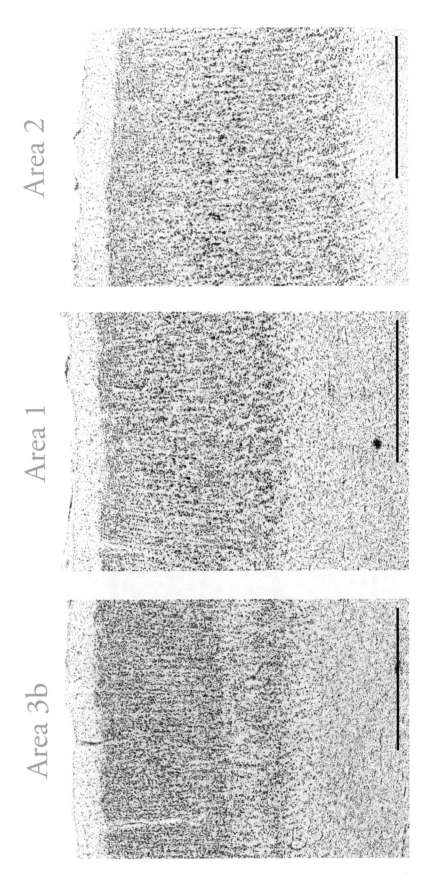

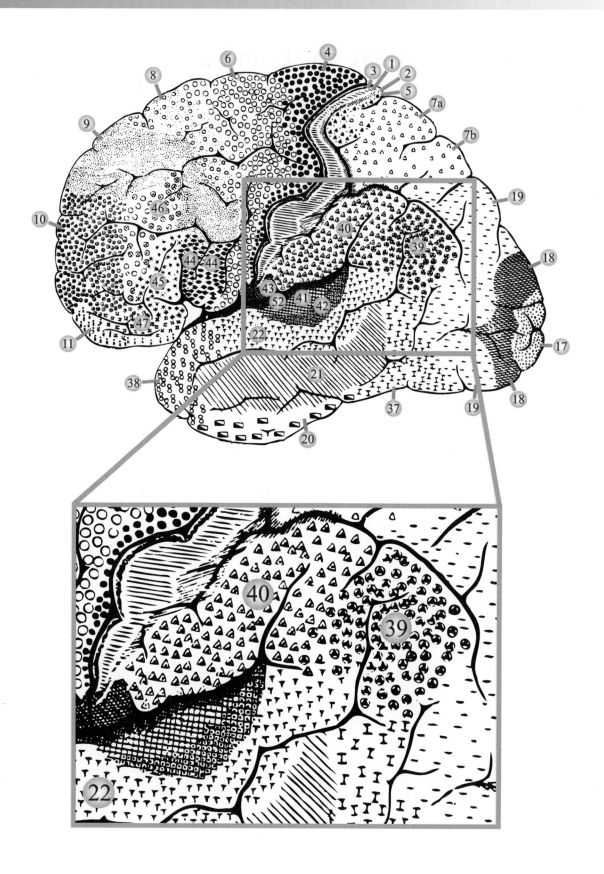

FIGURE 33 Brodmann's cytoarchitectonic map with the region of the inferior parietal lobule expanded (inset) to highlight the location of architectonic areas 40 and 39. Area 40 occupies the supramarginal gyrus and area 39 occupies the angular gyrus.

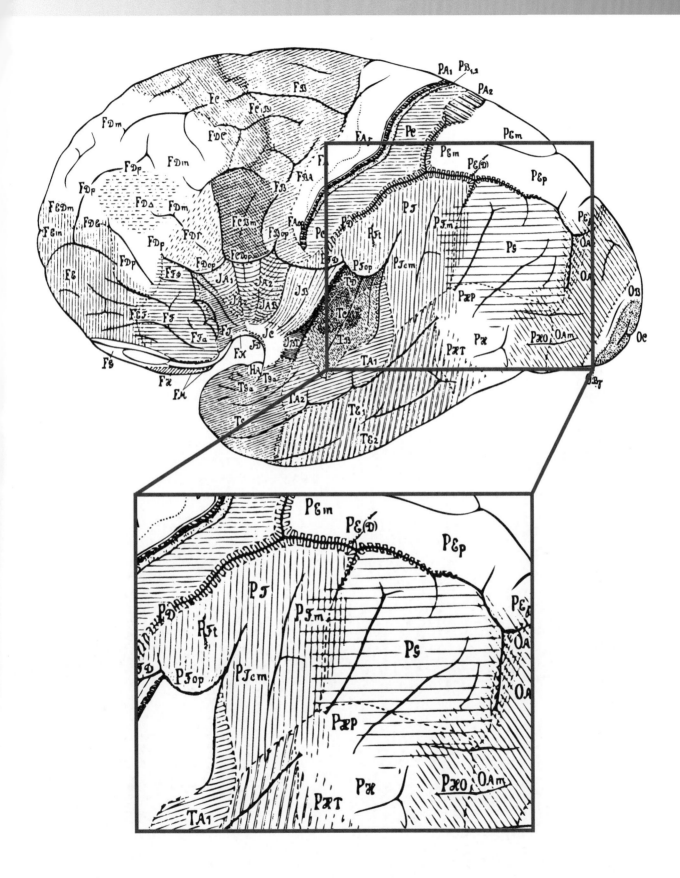

FIGURE 34 Cytoarchitectonic map of Economo and Koskinas (1925) with the region of the inferior parietal lobule expanded (inset) to highlight the architectonic areas that comprise this region.

Area PF

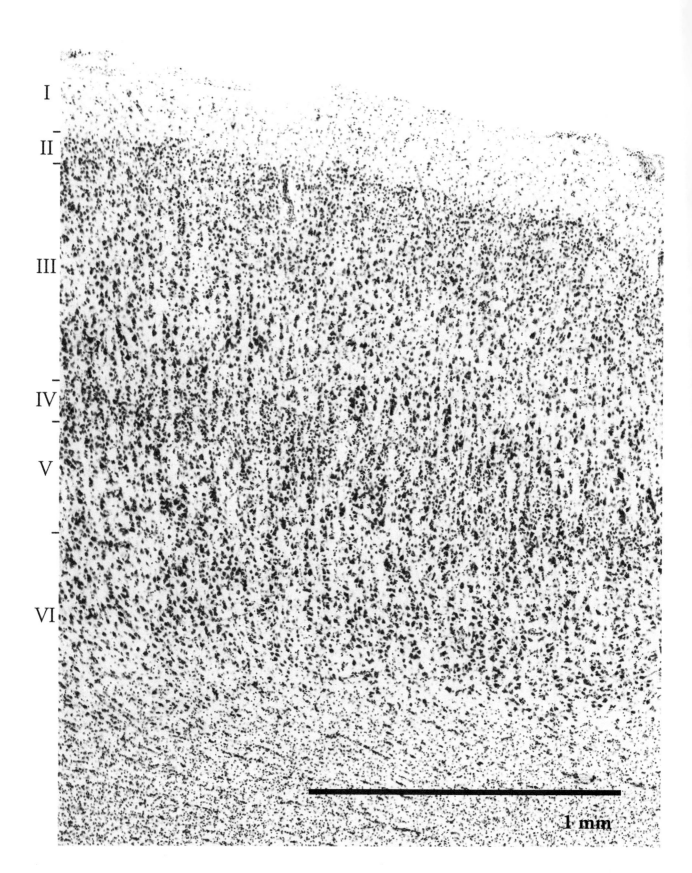

1 mm

Area PF

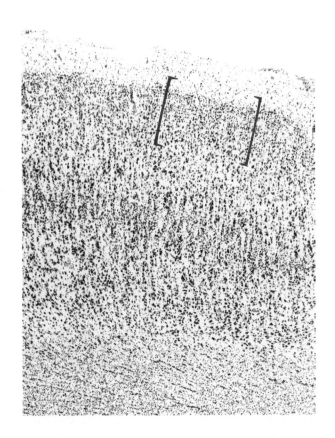

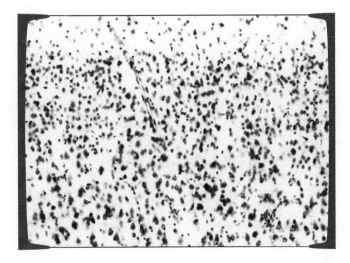

FIGURE 35 Photomicrograph of cytoarchitectonic area PF (anterior area 40) with a part of layers I, II, and III highlighted by the red brackets and expanded to illustrate the blurring of the border between layer II and the upper part of layer III.

INFERIOR PARIETAL LOBULE (SUPRAMARGINAL AND ANGULAR GYRI): AREAS PF, PFG AND PG

The cortex of the inferior parietal lobule was subdivided into two cytoarchitectonic areas by Brodmann: area 40 on the supramarginal gyrus and area 39 on the angular gyrus (Fig. 33). Economo and Koskinas (1925) referred to area 40 as area PF and area 39 as area PG (Fig. 34). Economo and Koskinas (1925), and later Sarkissov and colleagues (1955), recognized variations of the cortical type that covers the supramarginal gyrus. In recent years, the terminology of Economo and Koskinas (1925) has been frequently used for the areas of the inferior parietal lobule, partly under the influence of intensive research on the connections of their homologues in the macaque monkey (Pandya and Seltzer, 1982). In the macaque monkey, the rostral part of the inferior parietal lobule has been subdivided into area PF and area PFG. The homologue of area PG in the caudal part of the inferior parietal lobule (corresponding to the region of the angular gyrus) has been subdivided into area PG (anterior PG) and area Opt (probably corresponding to posterior PG).

In the following section, the essential features of the cytoarchitecture of three key areas of the inferior parietal lobule will be described with alternative names in parentheses. These are areas PF, PFG, and PG. Area PF is the cortex that occupies the most anterior part of the supramarginal gyrus and lies anterior and above the posterior ascending ramus of the lateral fissure (Fig. 18). This cortical area is the anterior part of area 40 of Brodmann and area PF of Economo and Koskinas (Figs. 33 and 34). Area PFG is the cortex that occupies the posterior part of the supramarginal gyrus that lies behind the posterior ascending ramus of the lateral fissure (Fig. 18). This part of the supramarginal gyrus has been included in area 40 by Brodmann and is the magnocellular variation PFcm of Economo and Koskinas (Fig. 34). Area PG, which occupies the central part of the angular gyrus, has been referred to as area 39 by Brodmann and as area PG by Economo and Koskinas (Figs. 33 and 34).

The nomenclature PF, PFG, and PG has been adopted to facilitate correspondence with the homologous areas in the macaque monkey in which the connectivity of these three basic areas of the inferior parietal cortex has been examined in detail: area PF is strongly connected with the ventral premotor area 6VR (where the orofacial musculature is represented) and has only weak connections with area 44, while area PFG has strong connections with area 44 and weak connections with area 6VR. Area PG is strongly connected with area 45, but not with area PF or PFG (Petrides and Pandya, 2009). There is also strong evidence that these connectional differences are preserved in the human brain (Kelly et al., 2010; Margulies and Petrides, 2013). For details, see next section (Connectivity of the Core Language Areas).

AREA PF (ANTERIOR PART OF AREA PF OF ECONOMO AND KOSKINAS, ANTERIOR AREA 40 OF BRODMANN)

The supramarginal gyrus can be divided into an anterior and a posterior part by the ascending posterior ramus of the lateral fissure (Fig. 18). Area PF lies on the anterior part of the supramarginal gyrus, caudal to the ventral part of somatosensory area 2. The difference between these two cortical areas is striking. In sharp contrast to the very large neurons encountered in the deepest part of layer III in area 2 (Fig. 31), the deepest part of layer III in area PF contains only medium size pyramidal neurons (photomicrograph of area PF on p. 124). As can be clearly seen in the photomicrograph, there is a gradient of neurons in layer III, ranging from small pyramidal neurons in its upper part to medium size neurons in its lower part. In area PF, the neurons in the uppermost part of layer III are very small and, therefore, they cannot be easily differentiated from the small neurons of layer II, giving the impression of an unusually broad layer II. The distinction between these two layers can be appreciated in the closer view of layer II afforded by figure 35. The pyramidal neurons in layer V are dispersed throughout the layer and thus a clear distinction between an upper and a lower part of layer of V cannot easily be made. In addition, layers V and VI cannot be easily distinguished because the medium-sized pyramidal neurons in layer V are not much bigger than those of layer VI.

Area PFG

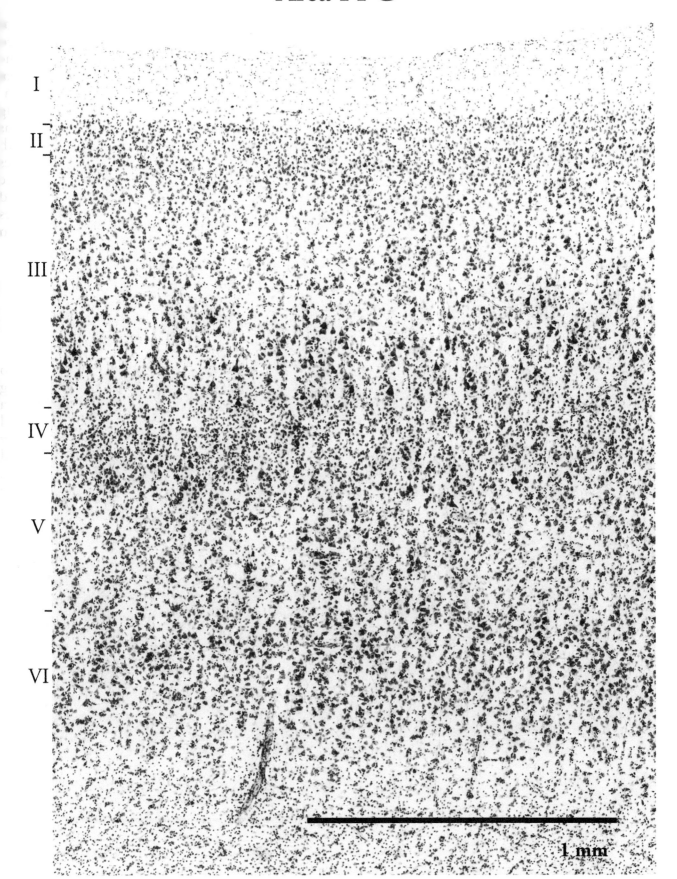

I

II

III

IV

V

VI

1 mm

Area PG

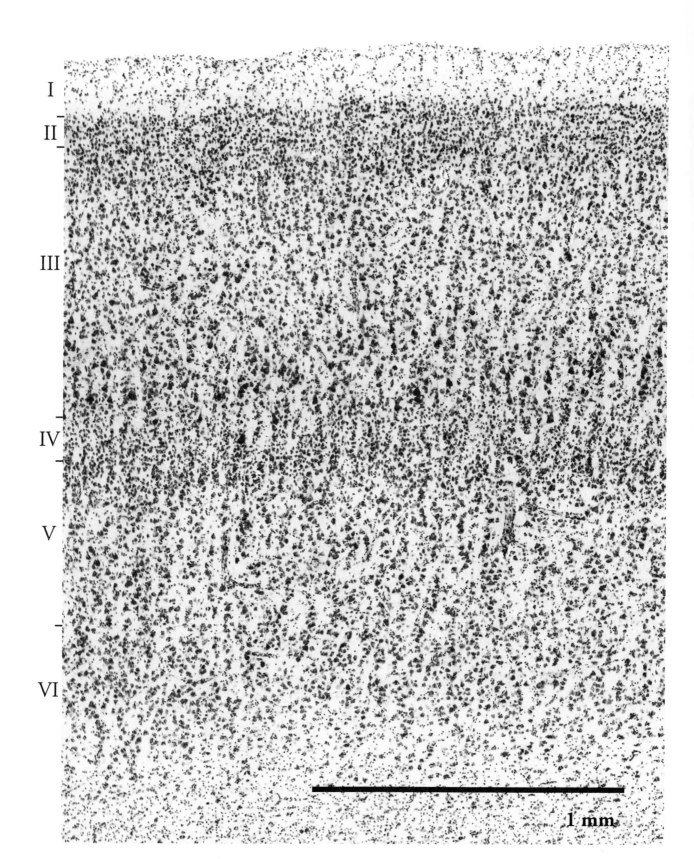

I

II

III

IV

V

VI

1 mm

Area PEm

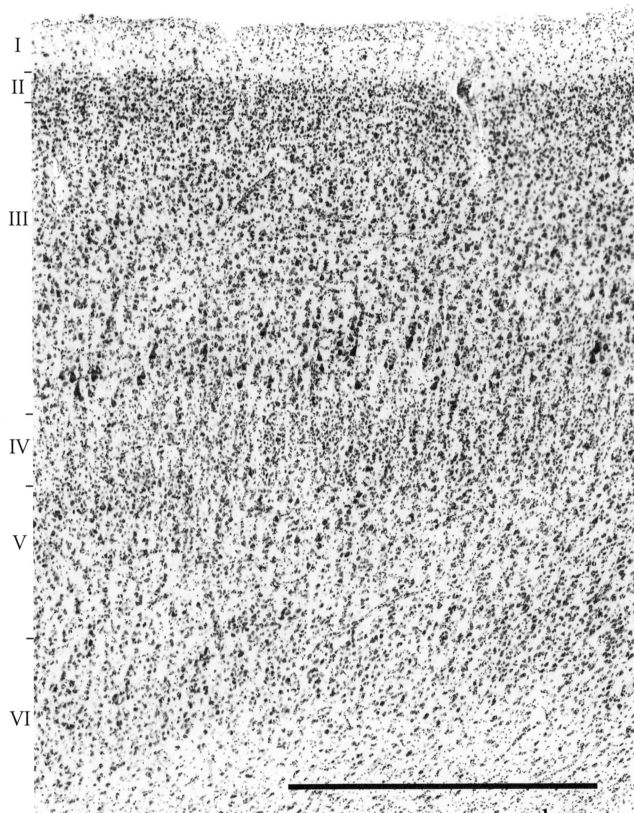

I
II
III
IV
V
VI

1 mm

Area PFG (posterior part of area PF of Economo and Koskinas, posterior area 40 of Brodmann)

Area PFG lies behind the posterior ascending ramus of the lateral fissure, on the posterior supramarginal gyrus (Fig. 18). At the parieto-temporal isthmus (x in Fig. 18), area PFG is succeeded, on the superior temporal gyrus, by the caudal part of area 22 of Brodmann, also known as area Tpt. The main features that differentiate area PFG from area PF are the somewhat better development of layer IV, the better differentiation of sublayers in III and V, and, on average, the slightly larger size of the neurons in the deeper part of layer III (compare photomicrographs on pages 124 and 127). It is the latter feature that led Economo and Koskinas (1925) to refer to this part of area PF as magnocellular (PFcm).

We have adopted the term "area PFG" to refer to the cortex of the posterior supramarginal gyrus because it corresponds to area PFG of the macaque monkey. In addition, it is the part of the inferior parietal lobule that is most strongly connected with area 44 in both the monkey (Petrides and Pandya, 2009) and the human brain (Margulies and Petrides, 2013).

Area PG (Area 39)

Caudal to the supramarginal gyrus lies the ill-defined region of the angular gyrus that stretches between the three caudal branches of the superior temporal sulcus which ascend into the inferior parietal lobule (Fig. 14). Brodmann (1909) labeled this region as area 39 and Economo and Koskinas (1925) as area PG (Figs. 33 and 34). In area PG, there is a clear differentiation of layer III into three sublayers. The gradient of very small pyramidal neurons in the uppermost part of layer III (just below layer II), the somewhat larger neurons in the central part of layer III and the even larger neurons in the deepest part of layer III just above layer IV is evident. In addition, layer V in area PG can be easily distinguished from layer VI because of the relative clearing (less densely packed neurons) in the deeper part of layer V. Recall that this distinction could only be made with difficulty in area PF because of the relatively even scattering of the

pyramidal neurons across layer V. Area PG occupie a very large region of the inferior parietal lobule an one can expect variants of it. One distinction tha has already been made is that between an anterio part, PGa, and a posterior part, PGp (Caspers e al., 2006, 2008).

Superior Parietal Lobule: Area PEm

Area PE is the cortex of the superior parietal lobe (Fig. 34). The anterior part of this large parietal cortical area, area PEm, has, in general, larger neurons (photomicrograph on page 129). Layer II is a thick dense layer of small neurons. There is a gradient of neuronal size in layer III, ranging from small to large neurons in the deepest part. Layer IV is thick and layer V is subdivided into an upper layer Va with medium-sized neurons and a sparse sublayer Vb with smaller and fewer neurons. The clearing in sublayer Vb is marked relative to that of the areas of the inferior parietal lobule.

The Superolateral Temporal Lobe (Superior and Middle Temporal Gyri): Areas TS3, Tpt, and 21

In the human and nonhuman primate brains, the processing of auditory information takes place in a series of cortical areas found on the lower bank of the lateral fissure and the adjacent superior temporal gyrus. The primary auditory cortex, which is found on the central part of Heschl's gyrus, has the typical koniocortical appearance of all primary sensory areas (e.g., Kaas and Hackett, 2000; Fullerton and Pandya, 2007). This core koniocortical area is surrounded by a number of other cortical areas, as is the case for the early stages of visual and somatosensory cortical processing (e.g., Merzenich and Brugge, 1973; Morel et al., 1993; Morosan et al., 2001; Rademacher et al., 2001; Petkov et al., 2006).

The critical role of the superolateral region of the temporal lobe in the dominant hemisphere for auditory processing of linguistic information was one of the earliest conceptual advances in the emerging understanding of the language regions of

the brain and figured prominently in the theoreti-
al writings of Wernicke (1874, 1881; see Fig. 4).
esion studies, electrical stimulation, and modern
unctional neuroimaging all agree that the central
nd posterior parts of the superior temporal gyrus,
he superior temporal sulcus, and the adjacent
ortex of the middle temporal gyrus are critical for
anguage processing (e.g., Penfield and Rasmussen,
950; Penfield and Roberts, 1959; Rasmussen
nd Milner, 1975; Dronkers et al., 1995; Binder
t al., 1997; Ojemann et al., 1989; Price, 2000,
010; Duffau, 2007, 2008; Friederici, 2011).
DeWitt and Rauschecker (2012) carried out a
meta-analysis of functional neuroimaging studies
of the human brain that has provided evidence for
an auditory word-form recognition stream in the
intermediate part of the superior temporal region
of the left hemisphere.

Brodmann (1909) referred to the central and
posterior parts of the superior temporal gyrus
as area 22 and the cortex of the adjacent middle
temporal gyrus as area 21 (Fig. 19). Economo and
Koskinas (1925) referred to the same regions as
area TA for the superior temporal gyrus and area
TE1 for the middle temporal gyrus (Fig. 20). The
cytoarchitecture of the superior temporal gyrus has
been examined in detail by Pandya and Sanides
(1973) in the macaque monkey and, more recently,
by Sweet and colleagues (2005) and Fullerton and
Pandya (2007) in the human brain. In this atlas,
the central part of Brodmann's superior temporal
area 22 has been labeled area TS3 and its posterior
part area Tpt (Fig. 18), in agreement with the corresponding homologous regions of the macaque
monkey (Fullerton and Pandya, 2007).

Area TS3 (anterior part of area 22 of Brodmann, TA of Economo and Koskinas)

Area TS3 lies on the superior temporal gyrus
somewhat anterior and lateral to the level of
Heschl's gyrus (Fig. 18). It is well known that the
koniocortical primary auditory region, which lies
on the central part of Heschl's gyrus, is sourrounded
by a belt of auditory areas that are extending onto
the adjacent dorsal part of the superior temporal
gyrus (see * in Fig. 18). Ventral to this region on
the superior temporal gyrus lies area TS3, which
represents one of the later stages of auditory analysis and which in the left hemisphere would be a
critical component of Wernicke's region. Indeed,
the location of the region that Wernicke (1881)
considered to be central to the auditory analysis of
language corresponds well with the location of area
TS3 and the immediately adjacent region in the
superior temporal sulcus (Fig. 4).

As can be observed in the photomicrograph of
area TS3 (p. 132), layer III contains pyramidal
neurons of small to medium size and the differentiation into sublayers is not clear. This becomes
evident when one compares layer III of area TS3
with that of an area such as PG which has a clear
differentiation of layers. Area TS3 has a distinct
layer IV. The pyramidal neurons in layer V are
small to medium size and are distributed throughout the layer. The distinction between layer V and
VI can be made with some difficulty.

Area Tpt

Caudally, area TS3 is succeeded by the tempo-
roparietal area Tpt which merges with the cortex
of the supramarginal gyrus (area PFG). As can be
seen in the photomicrograph of area Tpt (p. 133),
layer IV is broad and layers V and VI are composed
of small to medium-sized neurons. The distinction
between an upper and a lower part of layer V and
layer VI is difficult to make. Note the large neurons
in the deeper part of layer III.

Area 21

Area 21 occupies the central part of the middle
temporal gyrus. Layers V and VI are prevalent in
the cytoarchitectonic landscape of the middle temporal gyrus. The infragranular layers V and VI are
thicker and constitute a larger proportion of the
overall cortical thickness than the supragranular
layers II and III (photomicrograph on p. 134).
This characteristic differentiates middle temporal
cortical areas from inferior parietal areas. Layer V
is thick and contains several large neurons. By contrast, layer III is thin and occupies a much smaller
proportion of the overall cortical thickness.

Area TS3

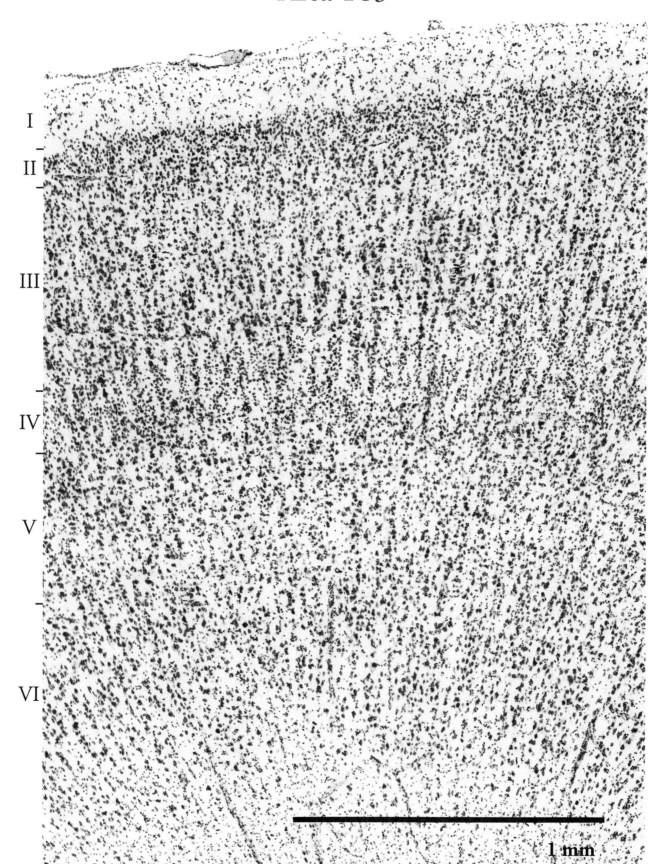

I

II

III

IV

V

VI

1 mm

Area Tpt

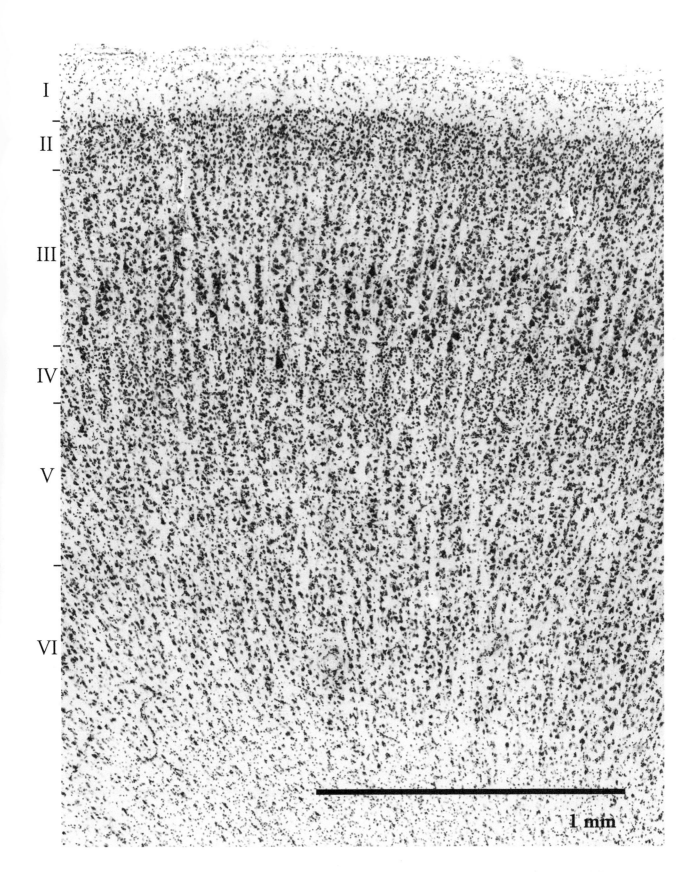

I

II

III

IV

V

VI

1 mm

Area 21 (MTG)

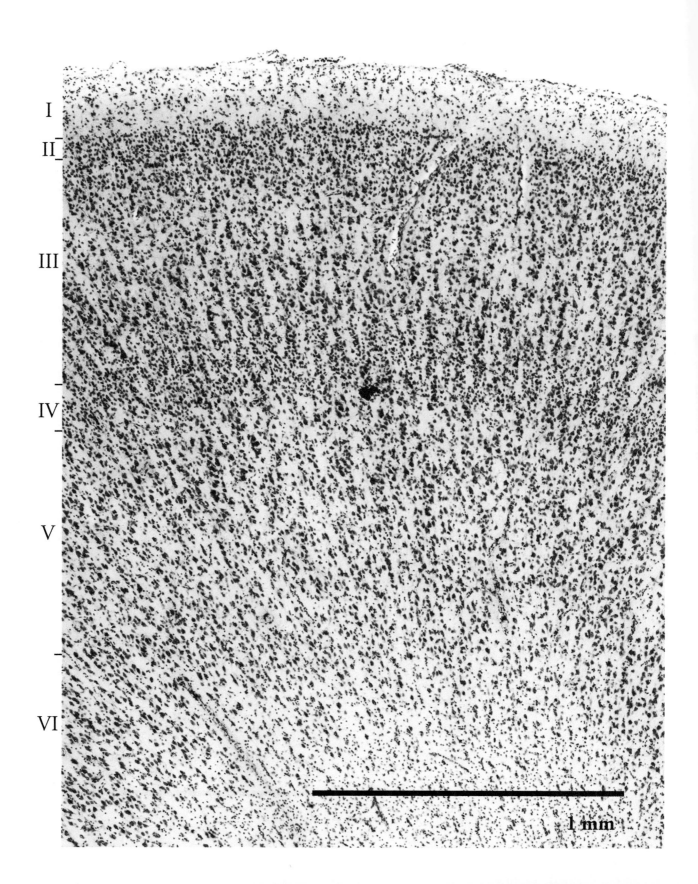

1 mm

Area TF

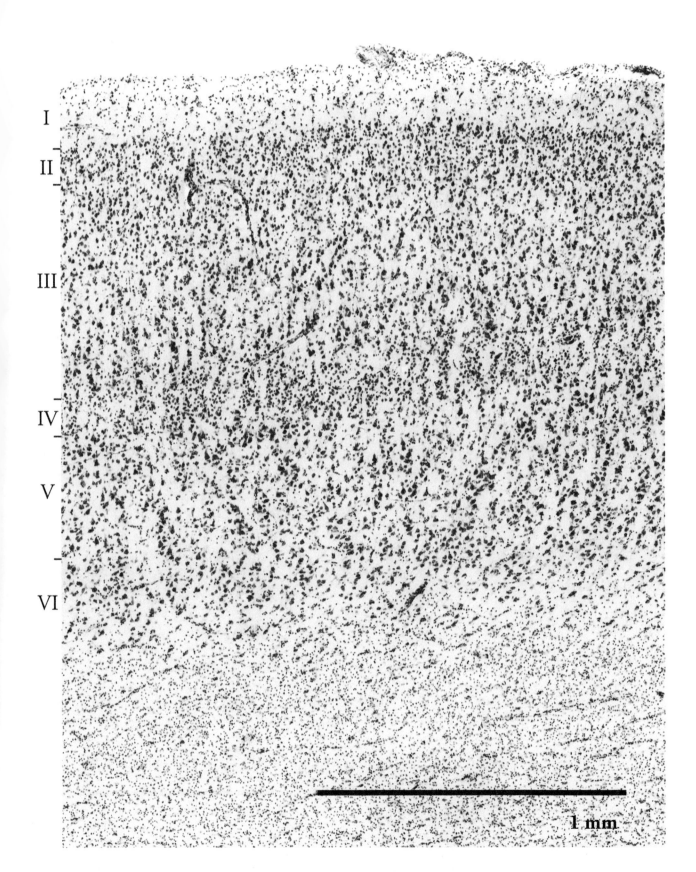

I

II

III

IV

V

VI

1 mm

Dysgranular Insula

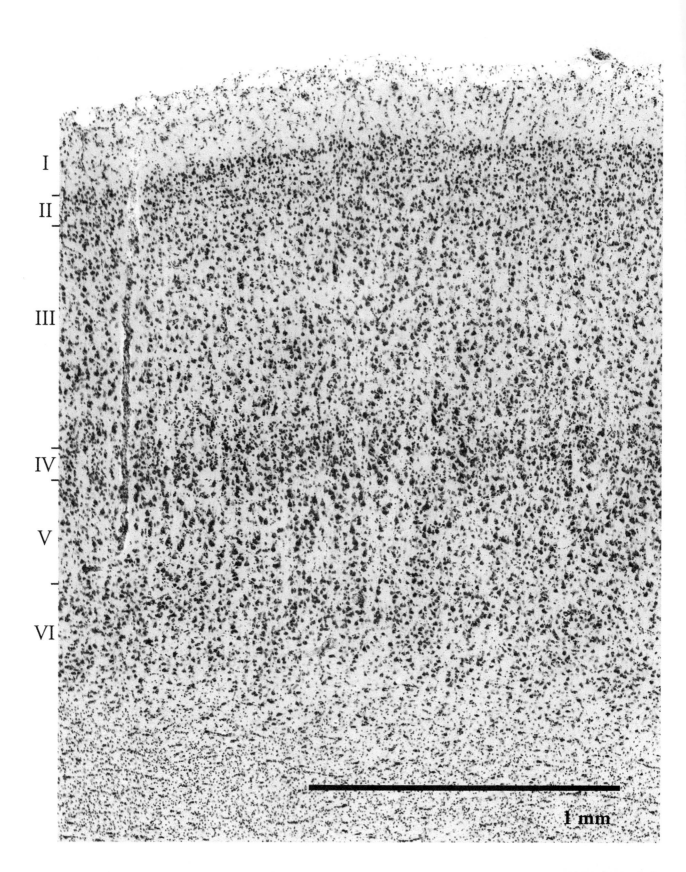

I

II

III

IV

V

VI

1 mm

Granular Insula

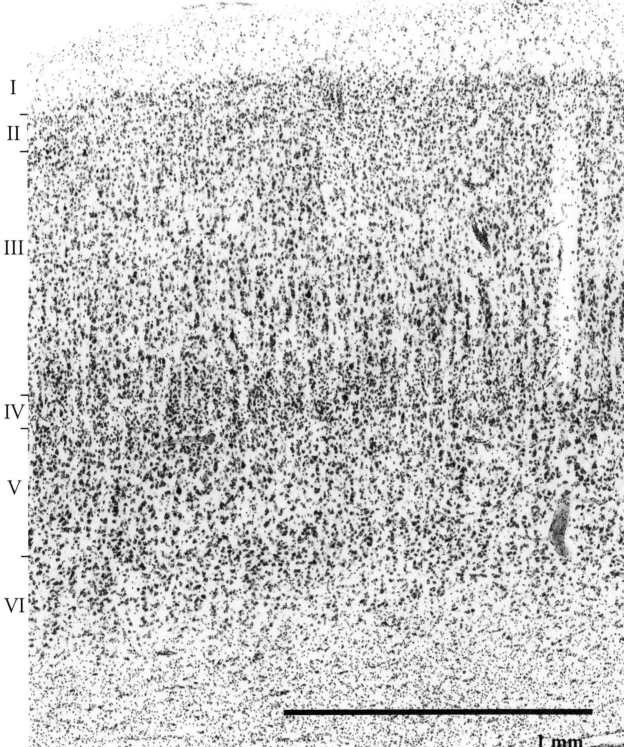

I

II

III

IV

V

VI

1 mm

The Fusiform Gyrus: Area TF

The cortical area TF, which is found on the fusiform gyrus, has a layer III with small to medium size neurons in its upper and lower parts, respectively (see photomicrograph on p. 135). Layer V is composed of groups of medium-sized pyramidal neurons in its upper part. The lower part of layer V contains medium-sized neurons scattered throughout the layer. The whole layer stains lightly and can easily be discriminated from layer IV, above, and layer VI, below.

The Insula: Dysgranular and Granular Areas

The anterior part of the insula, namely the three short gyri, is occupied by dysgranular cortex in which layer IV is a thin layer of small neurons interrupted at various points by the pyramidal neurons that lie above and below it (see photomicrograph on p. 136). The cytoarchitecture of this region is also characterized by large pyramidal neurons in the uppermost part of layer V, just below layer IV, which are disposed in a more or less horizontal row and known as the insular girdle. The lower part of layer V contains dispersed pyramidal neurons giving the appearance of an upper dense layer Va and a lower less dense layer Vb. Posteriorly, close to the central sulcus of the insula, the dysgranular cortex gradually changes to a granular cortical area in which layer IV is well developed (photomicrograph on p. 137). Thus, the two long gyri of the insula are occupied by a granular cortical area. The insular girdle is less evident in this granular cortical area. In the ventral part of the insular triangle, i.e., the insular pole, layer IV disappears and, therefore, the cortex is agranular. This agranular cortical area merges with the pyriform cortex at the orbito-temporal junction region and is part of the allocortical formations in the anterior medial temporal region.

Connectivity of the Core Language Areas

Connectivity of the Core Language Areas

Language processing is the result of the complex functional interactions between the core language areas and other cortical and subcortical structures. The classical model of language is based on two core language regions, namely Broca's region (for language production) and Wernicke's region (for comprehension of spoken language), and the connection between them. This model was originally proposed by Wernicke (1874, 1881) in the latter part of the 19th century and was subsequently elaborated in schematic diagrams by several investigators, the so-called "diagram makers" (e.g., Lichtheim, 1885). Although speculative linkages were frequently made between these and other areas to support theoretical ideas, remarkably little anatomical data were presented or discussed to support such connections.

Wernicke depicted a direct connection between the superior temporal region and the region of the inferior frontal gyrus across the lateral fissure (Fig. 36) and speculated that this connection may be made either directly via the white matter below the insula or via an intermediate link with the insula (Wernicke, 1881). Later, however, emphasis shifted to the arcuate fasciculus as the critical link between Wernicke's region and Broca's region (Geschwind, 1970) (Fig. 36). In both cases, real anatomical data were lacking, although Wernicke attempted (but failed) to dissect a pathway under the insula. Needless to say, detailed understanding of the connectivity of the core language areas is necessary for any sophisticated theoretical modeling of their role in language processes.

The classical method of examining the connections of cortical areas and subcortical structures has been the gross dissection of pathways in the white matter. The brain is first hardened in formalin or some other medium and the anatomist carefully dissects away the cortical grey matter before attempting to separate the white matter fascicles. This approach has provided a demonstration and definition of the classic association fiber pathways that are mentioned in standard neuroanatomy texts: the superior longitudinal fasciculus, the arcuate fasciculus, the superior and inferior occipito-frontal fasciculi, the uncinate fasciculus, the inferior longitudinal fasciculus, and the cingulate fasciculus (e.g., Curran, 1909; Klingler, 1935; Ludwig and Klingler, 1956; Klingler and Gloor, 1960; Fig. 37). The problem with this method is that while major fiber tracts running in a particular direction can be readily dissected, the precise origin and termination of the axons that form these fasciculi cannot be easily demonstrated. In other words, it is difficult to determine whether the fasciculus that has been isolated consists of monosynaptic connections (i.e. axons from area X to area Y) or whether it also includes fibers from other areas that mingle with these monosynaptic axons, but are directed to different parts of the brain.

Figure 37 illustrates the inferior occipito-frontal fasciculus, a major pathway of fibers running from the occipital region across the temporal lobe

---

FIGURE 36 Wernicke's (1881) and Geschwind's (1970) schematic diagrams illustrating the fundamental connection between the language comprehension zone in the superior temporal gyrus and the language production zone in the posterior part of the inferior frontal gyrus (Broca's region). Wernicke depicts the connection as a direct link across the lateral fissure, whereas Geschwind depicts the fundamental connection via the arcuate fasciculus, a set of fibers arching around the end of the lateral fissure. Abbreviations: a, auditory pathway; A, arcuate fasciculus; B, Broca's region; c, central sulcus; e, parallel sulcus (superior temporal sulcus); fS, fissure of Sylvius (lateral fissure); g, inferior occipital sulcus; i, intraparietal sulcus; k, anterior occipital sulcus; m, pathway to speech musculature; o, parieto-occipital fissure; x, sensory speech center; xy, association bundle between the two centers; y, motor speech center; W, Wernicke's area; I, inferior frontal gyrus; II, middle frontal gyrus; III, superior frontal gyrus. From Wernicke, C. (1881, p. 205) and Geschwind (1970), with permission. Color added to the original figures.

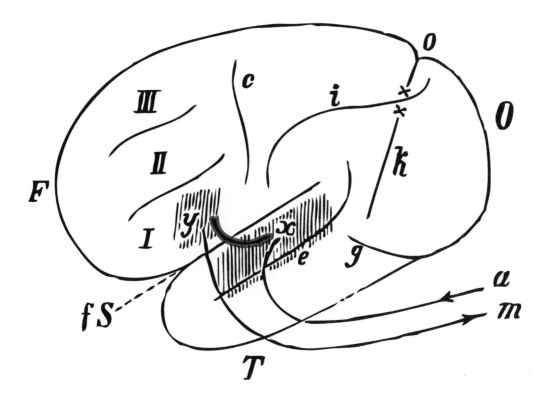

Wernicke

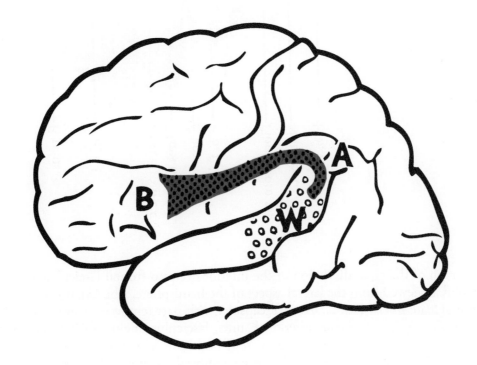

Geschwind

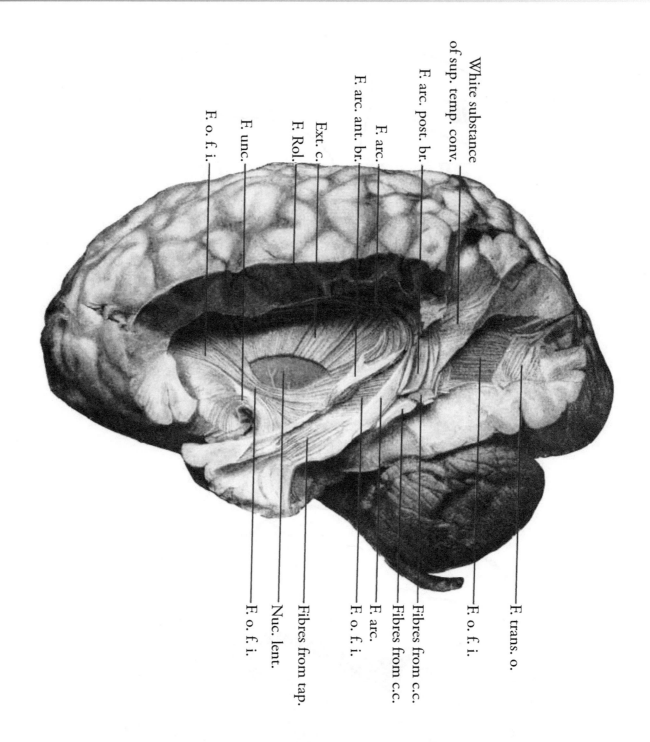

FIGURE 37 Photograph of the gross dissection of the inferior occipito-frontal fasciculus by Curran (1909). The course of this fasciculus as viewed from the lateral aspect of the hemisphere. Abbreviations: c.c., fibers from corpus callosum mixed with arcuate fibers; Ext. c., external capsule, with part removed to show Nuc. lent., nucleus lentiformis; F.o.f.i., fasciculus occipito-frontalis inferior; F. unc., fasciculus uncinatus; F. Rol., fissure of Rolando; F. arc., fasciculus arcuatus; F. arc. ant. br., anterior branch of fasciculus arcuatus; F. trans. o., posterior part of the fasciculus transversus occipitalis, most of which has been removed to show a large window through which the inferior occipito-frontal fasciculus can be observed; tap., fibers from the tapetum perforating the corona radiata and the inferior occipito-frontal fasciculus. (Reprinted with permission of John Wiley and Sons, Curran, E. J. (1909), "A new association fiber tract in the cerebrum with remarks on the fiber tract dissection method of studying the brain", Journal of Comparitive Neurology, 19, 645-656.) Color added to the original figure.

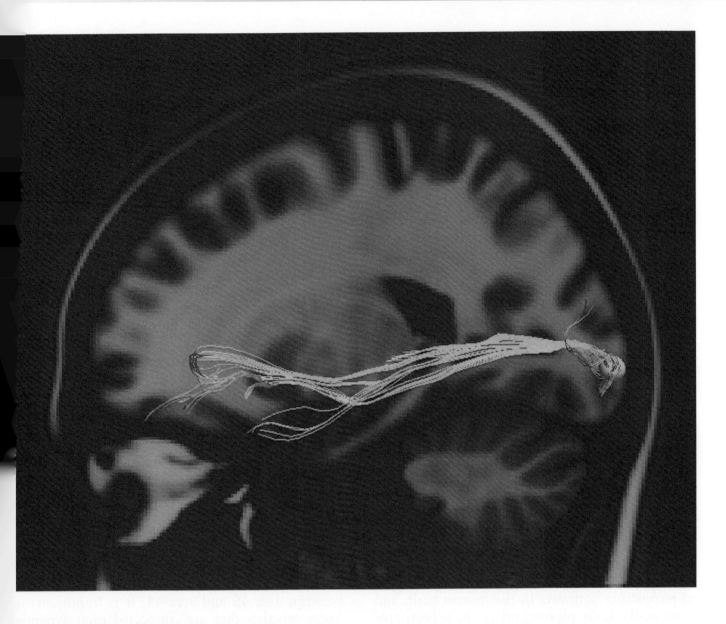

FIGURE 38 Diffusion MRI reconstruction of the inferior occipito-frontal fasciculus. This fasciculus is one of the easiest fasciculi to reconstruct with modern diffusion MRI. In the current reconstruction, we placed a seed in the occipital region in one brain and the fasciculus proceeding from the occipital region across the temporal lobe towards the frontal lobe was easily reconstructed. The diffusion data were acquired with a 3T MRI scanner (Philips Gyroscan superconducting magnet system) using an A-P phase sequence. The voxel size was 2mm³ and 64 slices were acquired. The tracks were placed in MNI space by registering them to the standard ICBM 152 model with Diffusion Toolkit and TrackVis. A 4.25mm sphere was used as the region of interest seed, which can be seen in the occipital lobe.

towards the frontal lobe. Although it can be readily demonstrated in gross dissections of the white matter of the temporal lobe, it is not possible to know how many of the fibers actually originate from the occipital lobe (and more specifically from which of the many visual areas within the occipital lobe) and are directed monosynaptically to the frontal lobe and how many of the fibers in the course of this fasciculus are axons from various temporal cortical areas that join those coming from the occipital cortex. Curran (1909), a master in the classical method of gross dissection who first identified this fasciculus, pointed out this problem.

In recent years, diffusion magnetic resonance

imaging (MRI), such as diffusion tensor imaging (DTI) and diffusion weighted imaging (DWI), has emerged as a major method of reconstructing in vivo the axonal pathways that link different regions of the human brain. Several articles have now been published providing reconstructions of the major pathways and a number of these have focused on reconstructions of those pathways that are particularly relevant to language processing (e.g., Catani et al., 2005; Croxson et al., 2005; Makris et al., 2005, 2009; Parker et al., 2005; Anwander et al., 2007; Vernooij et al., 2007; Catani and Mesulam, 2008; Catani and Thiebaut de Schotten, 2008; Frey et al., 2008; Glasser and Rilling, 2008; Saur et al., 2008; Kaplan et al., 2010). Careful examination of these reconstructions shows that, in most cases, although the core of the major fasciculi can be identified with some degree of accuracy, the precise origins and terminations of the reconstructed pathways cannot be provided (Figs. 38-40). This is due to the inherent limitations of current diffusion MRI tractography. For a review of these limitations in reconstructing the language pathways, see Campbell and Pike (2013).

Diffusion magnetic resonance imaging (MRI) records the motion of water molecules in the brain and, based on this information, the microstructure of the brain tissue around which the water is moving is inferred. Although diffusion MRI tractography has become popular as a method to reconstruct pathways in the human brain and some aesthetically pleasing images have been produced, it is important to understand the current limitations of the method that can lead to major errors in the reconstruction of these pathways (see, Hagmann et al., 2006; Johansen-Berg and Behrens, 2006; Jones, 2010; Jones and Cercignani, 2010; Jbabdi and Johansen-Berg, 2011; Jones et al., 2013; Campbell and Pike, 2013). With the current methodology, false positives and false negatives abound. It is unwise, at present, for an investigator of the neural bases of language to make strong claims and base theoretical arguments on connectivity inferred from diffusion MRI data. Methodological improvements of diffusion MRI and other promising methods (e.g., Axer et al., 2011) may make it possible to reconstruct the human pathways with accuracy in the future, but this is not the case at present. The comments made

above apply to its use in reconstructing the precise origin, pathway, and terminations of connection of the cortical areas of the brain. It should be noted that diffusion MRI can be very useful in clinical practice, such as in the evaluation of fiber integrity in various neurological conditions.

At this point, it becomes necessary to clarify what we mean by a "connection" between cortical areas because the term has been used ambiguously in the language literature. The cytoarchitectonically defined cortical areas are functional units of the brain that compute information and interact with specific distant cortical and subcortical areas via long axons. Thus, the question of interest is with which cortical and subcortical areas does area A interact monosynaptically? There is no doubt that all areas of the brain are connected with all others via multiple intermediate steps. For instance, we can say that area A is linked via intermediate steps with area D because area A is connected with area B, area B is connected with area C, and, finally, area C is connected with area D. The fundamental question, however, is whether axons of neurons in area A terminate directly upon neurons in area D and vice versa, forming monosynaptic connections which would indicate that the computation of information within area A is directly affecting what is occurring in area D. Here is an example directly relevant to language processing. If we are searching to understand the functional difference between area 45 and area 44, it is important to know whether they are connected monosynaptically with the same or different cytoarchitectonic areas of the parietal and temporal cortex. If we wish to develop a hypothesis/model about the role of area 45 in language, we wish to know whether it has direct access to the orofacial part of the motor cortex or whether interaction with motor areas is made, indirectly, via some other area(s) (see Petrides, 2006).

In the macaque monkey, these questions about monosynaptic connections can be readily answered with anterograde and retrograde anatomical tracers, the gold standard methods to examine anatomical connectivity. Anterograde tracers injected within a particular cortical area label the axons originating from neurons in that area and the entire course of these labeled axons can be followed to their final destination even within particular layers of specific

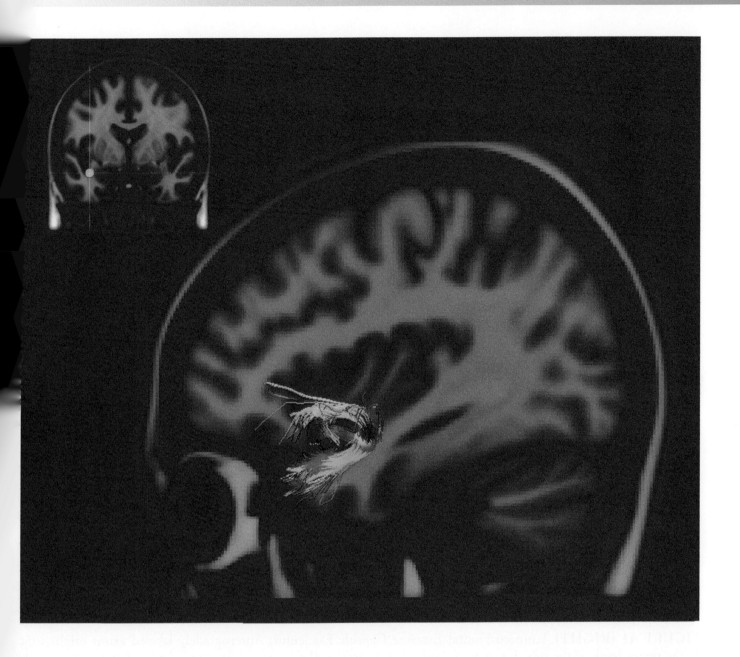

FIGURE 39 Diffusion MRI reconstruction of the uncinate fasciculus. This fasciculus links the anteriormost part of the temporal lobe with the orbital frontal region. In the current reconstruction, we placed a seed in the white matter close to the amygdala in one brain. The placement of the seed is indicated on the coronal section shown at the top left side of the figure. The sharply turning axons between the anteriormost part of the medial temporal region and the orbital/ventral frontal region can easily be visualized. It is not possible, however, to demonstrate the precise set of anterior temporal areas that project monosynaptically via this fasciculus and their exact terminations within particular orbital frontal cytoarchitectonic areas. Diffusion MRI details, as in figure 38.

distant cortical areas (e.g., Cowan et al., 1972; Fig. 41). Retrograde tracers injected into a particular cortical area are absorbed by axons of neurons that originate in distant areas and terminate in the injected area. The tracers are transported back to the neurons of origin and, therefore, label them, enabling the investigator to identify the precise cortical areas that project to the area within which the tracer was injected (e.g., Petrides and Pandya, 2002). Thus, in the experimental anatomical

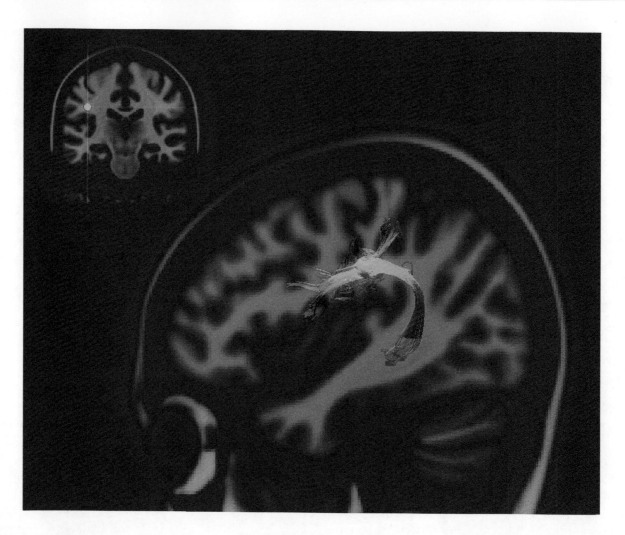

FIGURE 40 (ABOVE) Diffusion MRI reconstruction of the arcuate fasciculus. This fasciculus, which links posterior temporal and parietal cortex below the end of the lateral fissure with lateral frontal cortex, can be easily reconstructed by placing a seed close to the end of the lateral fissure. The placement of the seed is indicated on the coronal section shown at the top left side of the figure. Diffusion MRI details, as in figure 38.

FIGURE 41 (RIGHT) Temporo-Frontal Extreme Capsule Fasciculus. Anterogradely labeled axons originating in the upper bank of the superior temporal sulcus and entering the extreme capsule fasciculus demonstrated with the autoradiographic method in the macaque monkey brain. (A) Dark-field photomicrograph of a part of the ventrolateral prefrontal cortex to show terminations of axons. The inset shows the deep part of layer III and layers IV and V in this small part of cortex. The inset is expanded on the right side and photographed under light-field to show that the terminations of the axons are in a part of ventrolateral prefrontal cortex that has large neurons in deep layer III and a well developed layer IV, i.e., in area 45A. (B) Photomicrograph of the injection site of radioactively labeled amino-acids in the upper bank of the superior temporal sulcus to show the origin of the extreme capsule fasciculus (ECF) in this case. Note that the labeled axons course dorsally between the insula (IN) and the claustrum (CL) to enter the extreme capsule and continue towards the frontal lobe. This arrangement was typical of all cases with temporal lobe injections that demonstrated the ECF. Note that another branch of labeled fibers from the injection site is directed ventrally to other parts of the temporal lobe and a central branch is directed medially towards the thalamus and other medially located structures. Each one of the rectangles that constitute the overall photomicrograph is 2.955 mm by 2.205 mm and were taken by means of a motorized XY microscope stage and a computer running Stereo Investigator software (Microbrightfield, Inc.). Abbreviations: CL, claustrum; E, external capsule; ECF, extreme capsule fasciculus; IN, insula; LF, lateral fissure; LV, lateral ventricle; Pu, putamen; STS, superior temporal sulcus. (From Petrides and Pandya, 2009.)

A

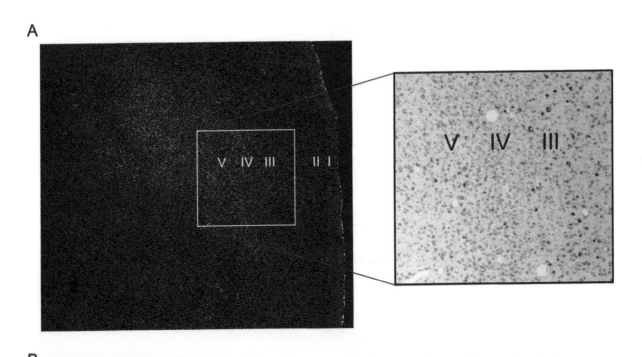

B

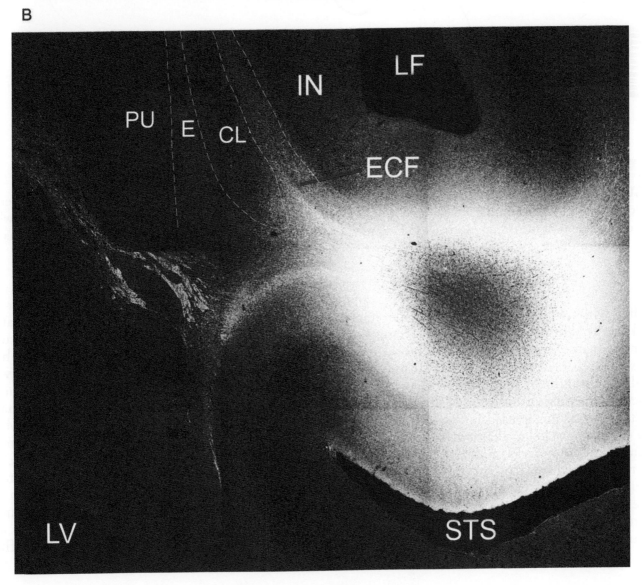

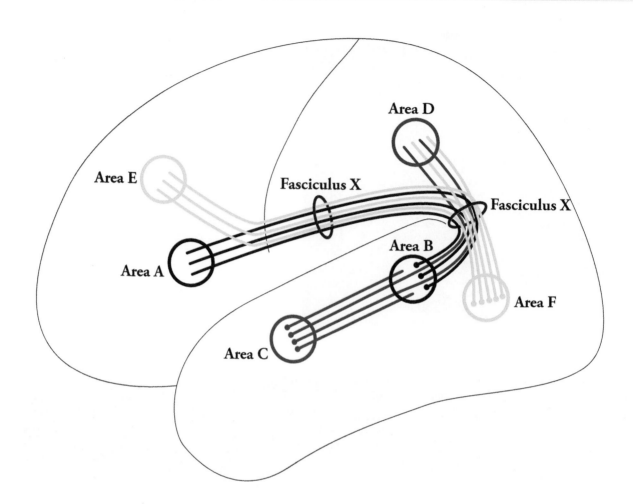

FIGURE 42 Conceptual schematic diagram to illustrate the problem in the definition of a fasciculus. Hypothetical fasciculus X is modelled on the arcuate fasciculus, an arching bundle at the crossroads of many axons in the white matter at the end of the lateral fissure. A to F represent hypothetical areas and the lines represent the monosynaptic links between them.

studies with anterograde and retrograde tracers used in the monkey, cortico-cortical connections refer to precise monosynaptic links between two areas. However, in the study of the classical fasciculi of the human brain with either the gross dissection method or diffusion MRI tractography, the term fasciculus is used rather loosely to refer to a collection of axons that enter a tract and may connect different areas monosynaptically or via several intermediate steps.

Figure 42 provides a schematic diagram to illustrate the important distinction between the precise connectivity diagram that we require for theoretical arguments and modeling of language processes and the concept of a fasciculus as a rather loosely

defined entity that contains multiple axons that do not necessarily link the same areas. Let us assume that with the gross dissection method we have isolated a bundle of axons, fasciculus X, and with diffusion MRI tractography we have reconstructed this fasciculus in vivo with more or less reasonable accuracy. We are interested in the question of which precise area or areas send monosynaptic axons to terminate in target area A via this fasciculus X. We discover that only area B sends such axons to area A (axons labeled in blue) and that other axons in this fasciculus are linking completely different areas: area F with area E (axons labeled in green) and area C with area D (axons labeled in red). Thus, fasciculus X may be a set of unrelated axons that

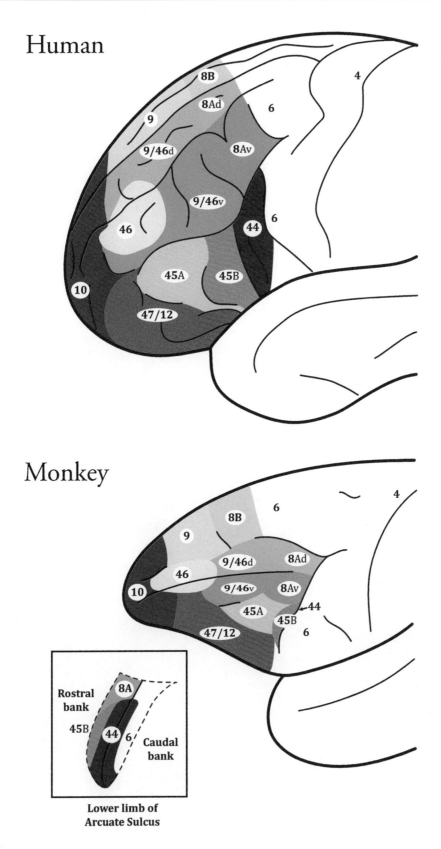

FIGURE 43 Cytoarchitectonic map of the frontal cortex of the human and macaque monkey by Petrides and Pandya (1994). Areas 44 and 45 in the macaque monkey were defined by the cytoarchitectonic criteria used to define these areas in the human ventrolateral frontal cortex. Area 44 in the macaque monkey lies in the fundus of the lower limb of the arcuate sulcus, part of which has been opened to show the location of this area (see inset).

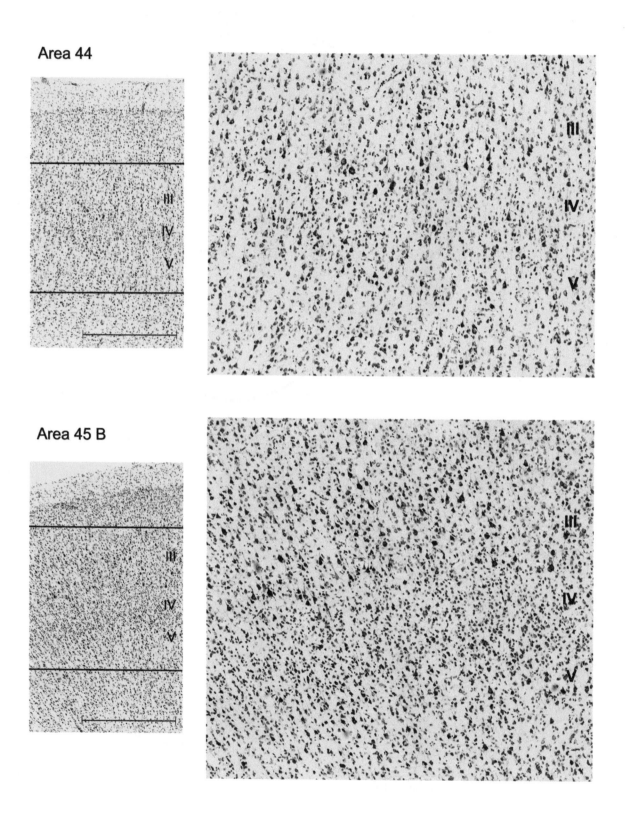

FIGURE 44 Photomicrographs of areas 44 and 45 in the macaque monkey. These areas were defined by the same cytoarchitectonic criteria used to define the corresponding areas of the human brain. (From Petrides and Pandya, 2009.)

ourse close to each other for a certain distance in certain part of the white matter. Furthermore, although gross dissection and current diffusion MRI can identify fasciculus X, they would have trouble discriminating the various axons that course through it and thus may erroneously lead to the conclusion that areas F and C send axons to area A via fasciculus X. Experimental anatomical tracing studies in macaque monkeys would have no problem discriminating the different types of connections depicted in figure 42 because tracers would be injected in each one of these areas and the precise course of the axons would be traced to their exact destination. Thus, information from macaque monkey experimental data is critical to evaluate claims made in studies of connectivity of the human brain with gross dissection or diffusion MRI.

A reasonable question at this point is whether precise studies of anatomical connectivity in the macaque monkey can contribute to our understanding of the anatomical connectivity of the core language areas, since macaque monkeys do not have language in the sense that we understand it in the human brain. We have addressed this issue in comparative cytoarchitectonic and connectivity studies of the macaque and human brains. We were able to establish, in the macaque monkey, the cytoarchitectonic homologues of areas 44 and 45 which, in the human brain, constitute Broca's region (Petrides and Pandya, 1994, 2002; Petrides et al., 2005) (Figs. 43 and 44). We then proceeded to establish the connections of these two cytoarchitectonic areas in the macaque monkey using anterograde and retrograde tracers (Petrides and Pandya, 2002, 2009; Frey et al., 2013). The cortico-cortical connection profiles of these two areas were quite distinct: area 44 had strong connections with area PFG of the inferior parietal lobule (the homologue of the posterior supramarginal gyrus in the human brain), but minimal if any connectivity with area PG in the caudal inferior parietal lobule (the homologue of the angular gyrus of the human brain). By contrast, area 45 had connections mainly with area PG. Another major difference between these two areas was their connectivity with the lateral temporal cortex. Whereas area 45 had massive connections with the cortex within the superior temporal sulcus and the cortex

immediately surrounding the superior temporal sulcus, area 44 had weak connectivity with these areas (Petrides and Pandya, 2009).

Thus, the macaque monkey data provided strong predictions about connectivity differences between areas 44 and 45 of the human brain. We proceeded to test these predictions using resting state functional connectivity (Kelly et al., 2010; Margulies and Petrides, 2013). This method indicates connections between cortical regions by demonstrating correlated activity across them and, at least partly, these functional connections are known to reflect anatomical connectivity (e.g., Biswal et al., 1995; Margulies et al., 2009). We were able to demonstrate strong connectivity differences between the two areas that comprise Broca's region, areas 44 and 45, in the human brain consistent with those predicted from the macaque monkey data: area 44 was shown to be functionally connected with area PFG in the posterior supramarginal gyrus, but not with area PG of the angular gyrus (Fig. 45), while area 45 was functionally connected with area PG, but not with area PFG (Fig. 46). In the temporal cortex, area 45 had strong connectivity with the superior temporal sulcus and the immediately adjacent cortex, exactly as predicted from the macaque monkey data (Margulies and Petrides, 2013).

A number of issues are raised by such results. Why should connectivity of an area in the macaque monkey be so strongly predictive of the connectivity pattern of the homologous language relevant area of the human cerebral cortex? To answer this question, we must recall that cortical areas, such as 44, 45, PFG, and PG, are not only contributing to language, but also play a major role in non-linguistic processing, a point increasingly appreciated by the scientific community. A simple example should suffice to make the point. The left inferior parietal lobule of the human brain (i.e. areas PF, PFG, and PG) has been shown to be critical for reading and writing since the early studies of Dejerine (1891a, 1891b, 1892). Reading and writing are human skills that have emerged not more than 5,000 years ago and these skills were the privilege of a tiny proportion of the population until very recently. Even today, according to UNESCO Statistics, 16% of the adult population is illiterate and there are countries in which illiteracy is more than 50%, and yet all these humans possess areas PF, PFG

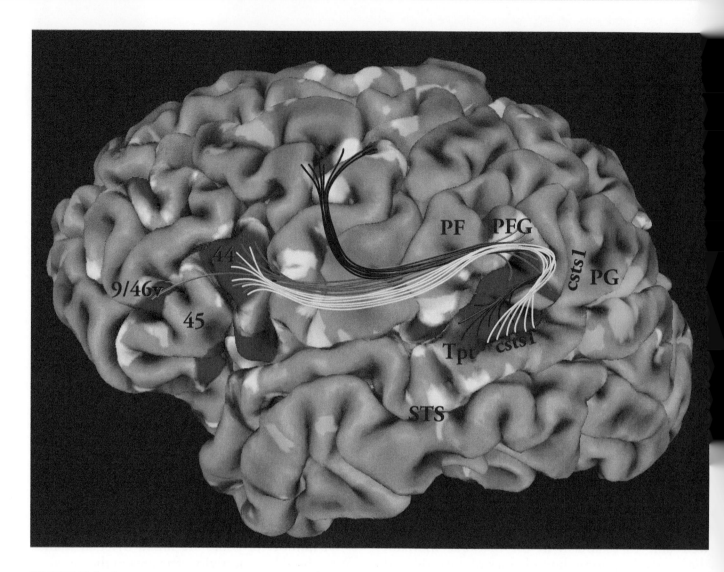

FIGURE 45 Resting state connectivity in the left cerebral hemisphere of one human brain showing positively correlated activity (indicated in red) between area 44 and the posterior part of the supramarginal gyrus (area PFG), as well as the adjacent posterior part of the superior temporal gyrus (area Tpt). Note the negatively correlated activity (indicated in blue) in the angular gyrus (area PG). For details, see Margulies and Petrides (2013). Superimposed on this image are the fasciculi known to connect these regions (yellow and blue, branches of arcuate fasciculus; green, third branch of the superior longitudinal fasciculus).

and PG in both the language dominant and non-dominant hemispheres. Clearly, the fundamental role of these areas in the human brain, including the left hemisphere in illiterate people, cannot lie in the processing of reading and writing, although the contribution of these areas is critical for the acquisition and maintenance of these skills in the human population. Indeed, we would argue that study of the non-linguistic contribution of these areas in the human brain and in the macaque monkey can provide insights into the reasons why

these areas are critical components of the network subserving reading and writing.

We must not confuse the issue of the fundamental connectivity of cortical areas, namely whether a particular area A is linked via monosynaptic axons with area B, but not with areas C and D, with other questions about connectivity, such as the relative strength of connections, hemispheric asymmetry in strength of connections, etc. The fundamental connectivity between cortical areas is likely to have been largely preserved between human and

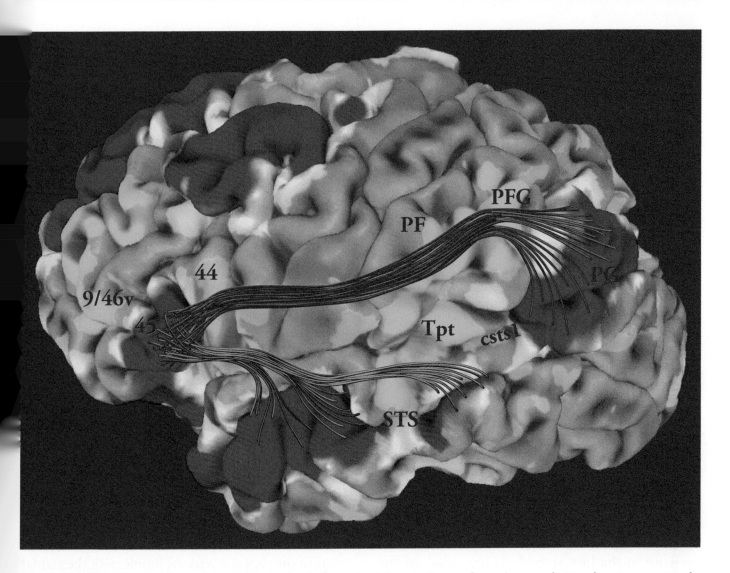

FIGURE 46 Resting state connectivity in the left cerebral hemisphere of one human brain showing positively correlated activity (indicated in red) between area 45 and area PG in the angular gyrus, as well as the superior temporal sulcus and adjacent cortex on the superior and middle temporal gyri. Note the negatively correlated activity (indicated in blue) in the posterior supramarginal gyrus (area PFG) and the posterior part of the superior temporal gyrus (area Tpt). For details, see Margulies and Petrides (2013). Superimposed on this image are the fasciculi known to connect these regions (dark purple, second branch of the superior longitudinal fasciculus; red, arcuate fasciculus; light purple, temporo-frontal extreme capsule fasciculus).

nonhuman primate brains for homologous areas, although there may be differences in the strength or hemispheric asymmetry of particular connections. For instance, the strength of the connections between particular areas may have increased in the human brain and it may be possible to link such differences to particular abilities, although one must always be cognizant of the inherent difficulties in accurately measuring such connections in the human brain (e.g., Campbell and Pike, 2013). The important point is that, if the fundamental

connectivity of cortical areas has been largely preserved between macaque monkey and the human brain, as suggested by available data (e.g., Margulies et al., 2009), then the macaque monkey findings provide useful information about the connectivity patterns of the human cortical areas, including those that are centrally involved in language (e.g., Kelly et al., 2010; Margulies and Petrides, 2013).

In the presentation of the pathways linking language relevant areas in this atlas, a distinction has been made between 1) the precise monosynaptic cortico-cortical connections of the various areas and the fasciculus or fasciculi within which these monosynaptic axons course and 2) the fasciculi as larger more loosely defined entities that contain multiple fibers linking disparate sets of areas (see Fig. 42). The language researcher must realize that the monosynaptic cortico-cortical axonal connections of area B with area A via fasciculus X is not synonymous with fasciculus X because fasciculus X contains additional fibers that link area F with area E and area C with area D (Fig. 42). Knowledge of the disparate axons that are entering and leaving a fasciculus at a particular point may be clinically useful in evaluating the deficits from a lesion at a specific location in the white matter. But theoretical models and hypotheses about the cortical language circuits require precise knowledge of the monosynaptic connections between particular areas.

Given the above considerations, the following approach has been adopted in this atlas for the discussion of the various fasciculi that link core language areas of the human brain. First, the data from experimental anatomical studies in the macaque monkey are used to present established monosynaptic axonal cortico-cortical connections via particular fasciculi on a three dimensional MRI of the monkey brain. Then, the presumed connectivity in the human brain is presented on the average MRI of the human brain with discussion of gross dissection, diffusion MRI, and resting state connectivity data that provide, at least partial, support for the existence of these axonal connections in the human brain. It is the belief of the author that such a presentation of the data may help the reader navigate between monosynaptic connectivity of cortical areas that is reasonably well established (i.e. demonstrable with the gold standard methods in the macaque monkey and supported by diffusion MRI and resting state connectivity in the human brain) and polysynaptic connectivity via one or more intermediate steps. It can also help evaluate claims in the literature about connections in the human brain that may be false positives or false negatives of the current methods (see Campbell and Pike, 2013).

Note that all cortico-cortical connections are bi-directional. Thus, terms such as parieto-frontal or temporo-frontal are merely conventional ways of referring to these fasciculi and do not imply that the connections are uni-directional.

SUPERIOR LONGITUDINAL FASCICULUS (BRANCHES I, II, III): THE PARIETO-FRONTAL INTERACTION SYSTEM FOR LANGUAGE PROCESSING

In the older anatomical literature, the horizontally oriented cortico-cortical axonal system linking the posterior parietal region with the lateral frontal cortex (not just Broca's region in the ventrolateral frontal lobe) was referred to as the superior longitudinal fasciculus (Dejerine 1895) (see Fig. 47). Petrides and Pandya (1984) examined this axonal fiber system in the macaque monkey using the autoradiographic method for tracing axonal connections. This research demonstrated that the superior longitudinal fasciculus can be divided into three major branches that were named Superior Longitudinal Fasciculus I (SLF I), Superior Longitudinal Fasciculus II (SLF II), and Superior Longitudinal Fasciculus III (SLF III) (Petrides and Pandya, 1984).

The dorsal branch of the superior longitudinal fasciculus (SLF I) is a completely independent system of cortico-cortical axons that links various areas of the superior parietal lobule with the caudal dorsolateral and dorsomedial frontal regions, including the supplementary motor region and the cingulate motor areas (Petrides and Pandya, 1984) (Figs. 48 and 49). For the language researcher, this is a major association axonal system that links the superior parietal region with the supplementary speech region. This part of the parieto-frontal axonal system had not been previously described from classic gross dissections of the white matter. Diffusion MRI has recently provided evidence for the existence of this pathway in the human brain also (Makris et al., 2005; Rushworth et al., 2006). In a recent functional neuroimaging study, we were able to show increased activation in many of the caudal dorsolateral and dorsomedial frontal areas and superior parietal areas that are linked together via SLF I during the act of writing (Segal and Petrides, 2012b).

The second component of the superior longitudinal fasciculus (SLF II) originates from the caudal

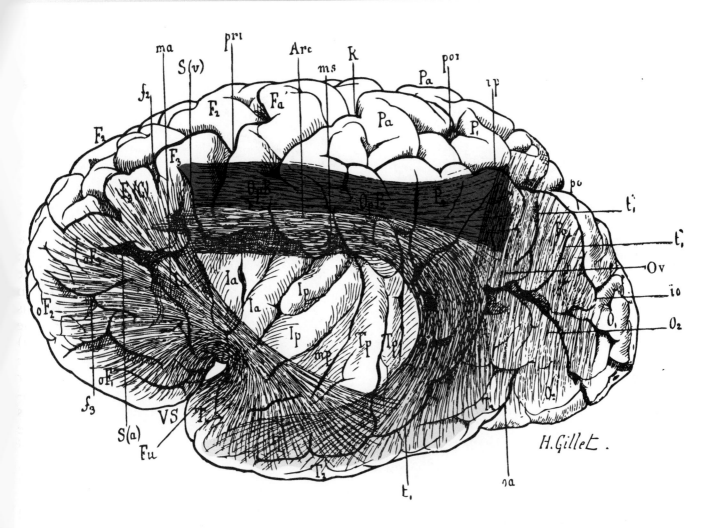

FIGURE 47 The schematic illustration of various fasciculi in the famous textbook of neuroanatomy by Dejerine (1895). The fibers colored in red are those of the superior longitudinal fasciculus running between the inferior parietal lobule and the frontal lobe. The fibers colored in yellow are those that are arching around the end of the lateral fissure linking temporal with parietal cortex and temporal with frontal cortex. In the discussion of these pathways, Dejerine considered the straight axons and the arching fibers to belong to the same system, in other words the arching fibers that we now consider to be part of the arcuate fasciculus were treated as part of the superior longitudinal fasciculus. Note that the first branch of the superior longitudinal fasciculus (SLF) later isolated in the macaque monkey by Petrides and Pandya (1984) and named SLF I is not shown in the schematic diagram of Dejerine. SLF I is a difficult fasciculus to isolate with gross dissection.

part of the inferior parietal lobule and links this region with various ventrolateral and dorsolateral frontal areas (Petrides and Pandya, 1984, 2009). The axons that link parietal area PG with the ventrolateral frontal area 45 course as part of this major axonal system (Petrides and Pandya, 2009) (Figs. 48 and 49). In the human brain, this is the fiber system that links the cortex of the angular gyrus with lateral frontal cortex. The axons that

originate from the lower parts of the angular gyrus are forced to curve around the end of the lateral fissure and can be considered to be part of the arcuate fasciculus, emphasizing the difficulty in making the distinction between some components of the arcuate fasciculus and SLF II (see discussion of the arcuate fasciculus below).

The third branch of the superior longitudinal fasciculus (SLF III) is a distinct component that

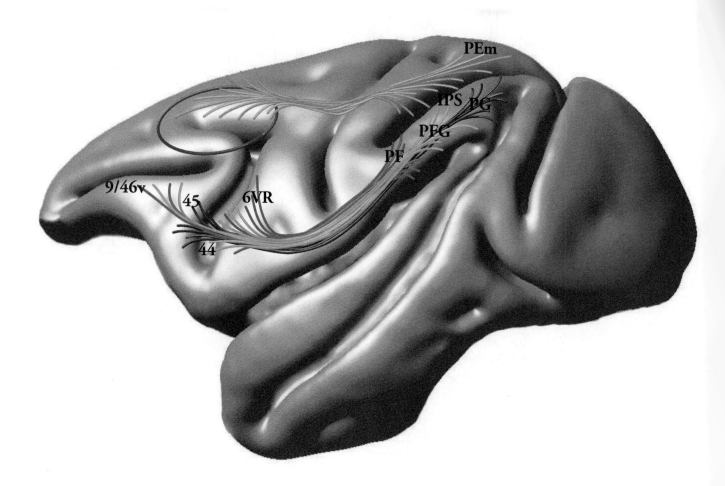

FIGURE 48 Schematic illustration of the three branches of the superior longitudinal fasciculus in the macaque monkey. The superior longitudinal fasciculus I (SLF I; orange) links the cortical areas of the superior parietal lobule with the caudal dorsolateral (outlined in red) and dorsomedial frontal cortex. SLF II (purple) links the caudal part of the inferior parietal lobule with ventrolateral frontal cortex, primarily area 45. SLF III (green) links the rostral part of the inferior parietal lobule with the rostroventral part of area 6 and area 44.

arises from the rostral part of the inferior parietal lobule (Petrides and Pandya, 1984, 2009), which in the human brain would be the morphological entity known as the supramarginal gyrus (Figs. 48 and 49). Diffusion MRI has provided evidence for the existence of this pathway in the human brain (Makris et al., 2005; Rushworth et al., 2006; Frey et al., 2008; Bernal and Altman, 2010). The axons originating from the rostralmost part of the inferior parietal lobule (area PF) target the ventral premotor cortex and to a lesser extent area 44,

while those originating further caudally from area PFG target primarily area 44 (Petrides and Pandya, 2009). Thus, SLF III is the main pathway linking the rostral part of the inferior parietal lobule (the supramarginal gyrus in the human brain) with area 6VR and area 44, while SLF II is the main axonal pathway that links the caudal part of the inferior parietal lobule (the angular gyrus in the human brain) with area 45 of the ventrolateral frontal cortex. Although the definitive evidence about these two branches of the superior longitudinal

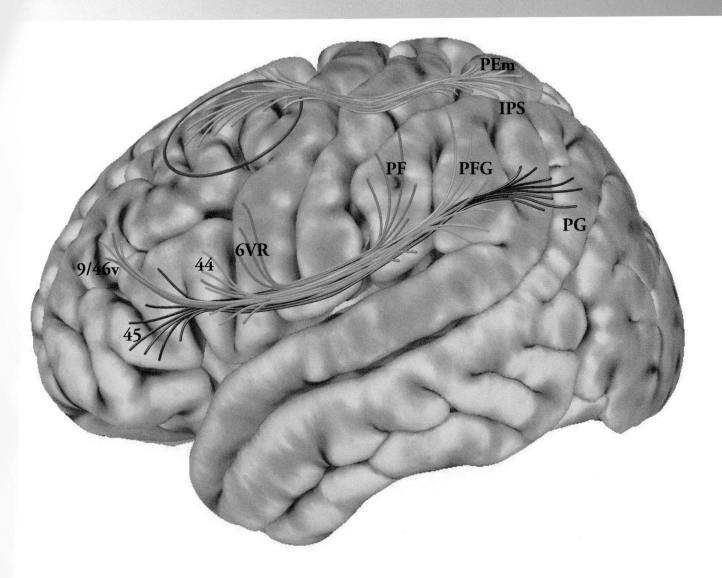

FIGURE 49 Schematic illustration of the three branches of the superior longitudinal fasciculus on the average MRI of the human brain. The superior longitudinal fasciculus I (SLF I; orange) links the cortical areas of the superior parietal lobule with the caudal dorsolateral (outlined in red) and dorsomedial frontal cortex. SLF II (purple) links the caudal part of the inferior parietal lobule, which in the human brain is the angular gyrus, with ventrolateral frontal cortex, primarily area 45. SLF III (green) links the rostral part of the inferior parietal lobule, which in the human brain is the supramarginal gyrus, with the rostroventral part of area 6 and area 44.

fasciculus, SLF II and SLF III, comes from autoradiographic studies in the macaque monkey (Petrides and Pandya, 1984, 2009), we have recently provided evidence with diffusion MRI (Frey et al., 2008) and resting state connectivity (Kelly et al., 2010; Margulies and Petrides, 2013) that the connections in the human cortex are consistent with those predicted by macaque monkey data.

THE ARCUATE FASCICULUS: THE CLASSIC LANGUAGE PATHWAY

The standard model of the neural basis of language is that of a superior temporal language region linked with an anterior ventrolateral frontal region by the arcuate fasciculus, a bundle of axons arching around the posterior end of the lateral fissure (see Catani and Mesulam, 2008). This model was popularized in the latter part of the 20th century by Geschwind (1970) (Fig. 36). However, no

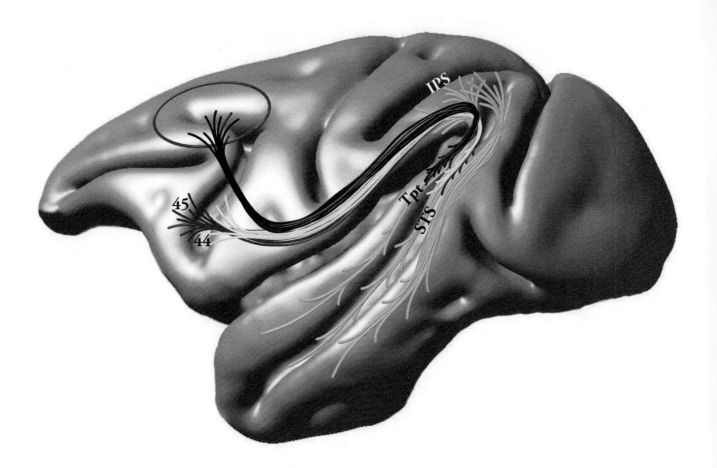

FIGURE 50 Schematic illustration of the arcuate fasciculus (dark blue, yellow, and red) and the middle longitudinal fasciculus (light blue) in the macaque monkey brain to illustrate the complex relation between these two pathways. The fibers of the arcuate fasciculus arching around the end of the lateral fissure link posterior temporal cortex with lateral frontal cortex. These fibers are arching around the end of the lateral fissure in the same region where fibers from anterior and middle temporal cortical areas are directed to the inferior parietal lobule, i.e. the middle longitudinal fasciculus. Thus, the end of the lateral fissure is the crossing point of many axonal fibers.

anatomical evidence was presented for the connection between the posterior third of the superior temporal gyrus and Broca's region (areas 44 and 45).

What is the anatomical evidence for the arcuate fasciculus? Arching fibers around the end of the lateral fissure, giving the impression of linking the lateral frontal cortex with the temporo-parietal junction region just below the end of the lateral

fissure, can be readily observed in gross dissections of the human white matter and these were discussed as early as 1895 by Dejerine (Fig. 47). Here it should be pointed out that the arcuate fasciculus was not always treated as separate from the superior longitudinal fasciculus (e.g., Dejerine, 1895). Indeed, it was thought to be nothing more than that component of the superior longitudinal fasciculus which originates from the posterior

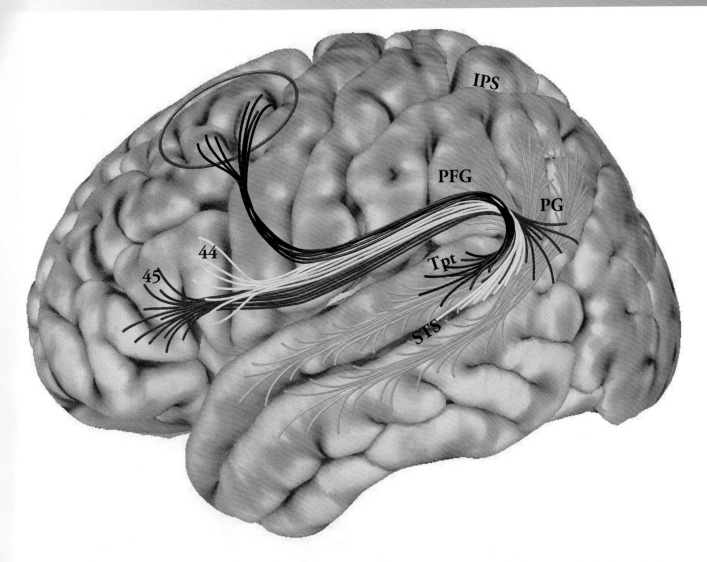

FIGURE 51 Schematic illustration of the arcuate fasciculus (dark blue, yellow, and red) and the middle longitudinal fasciculus (light blue) in the human brain to illustrate the complex relation between these two pathways. The fibers of the arcuate fasciculus arching around the end of the lateral fissure link posterior temporal cortex with lateral frontal cortex. These fibers are arching around the end of the lateral fissure in the same region of the white matter where fibers from anterior and middle temporal cortical areas are directed to the inferior parietal lobule, i.e. the middle longitudinal fasciculus.

temporo-parietal region below the end of the lateral fissure, thus forcing axons from this region to curve around the end of the lateral fissure on their way to the frontal cortex. Even today, reconstructions of the arcuate fasciculus are referred to as the "superior longitudinal/arcuate fasciculus", reflecting the difficulty in separating the axons of these two fasciculi as they course around the end of the lateral fissure towards the frontal lobe.

Although it is easy to dissect out and reconstruct with diffusion MRI an arching set of axons around the end of the lateral fissure (e.g., Catani and Thiebaut de Schotten, 2008; Glasser and Rilling, 2008; Rilling et al., 2008; Upadhyay et al., 2008; Fig. 40), it is difficult to know exactly which areas are linked monosynaptically via these arching fibers and separate these connections from those of the superior longitudinal fasciculus that

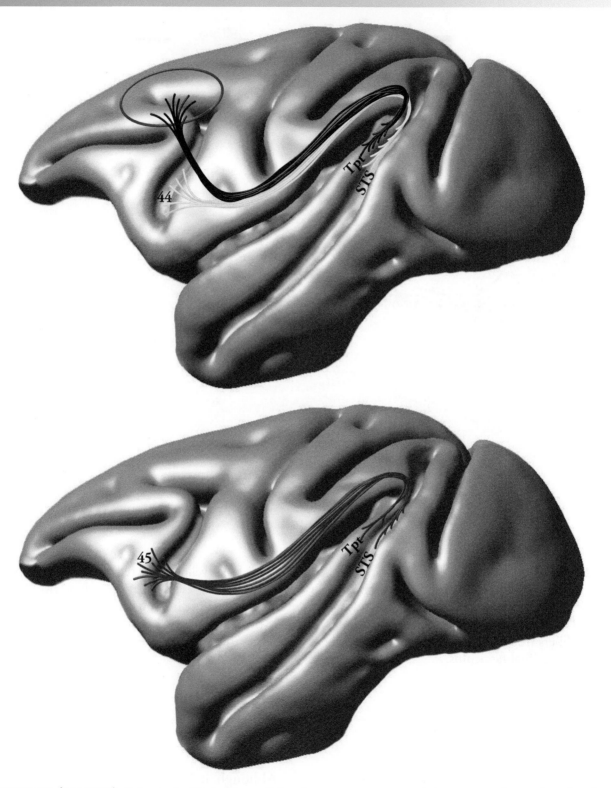

FIGURE 52 (ABOVE) Schematic illustration of the three components of the arcuate fasciculus that have been isolated in the macaque monkey brain with experimental tracer methods. Dark blue: branch originating from the dorsal part of the caudal superior temporal gyrus (close to the lateral fissure) and directed to the posterior dorsolateral frontal cortex outlined in red (area 8Ad and area 6DR). Yellow: branch originating from the ventralmost part of the gyrus and the cortex in the upper bank of the superior temporal sulcus and terminating in area 44. Dark red: branch originating from the caudal superior temporal gyrus and the superior temporal sulcus terminating in area 45. **(RIGHT)** Schematic illustration of the three components of the arcuate fasciculus that have been isolated in the macaque monkey brain with experimental tracer methods projected onto the human average MRI brain.

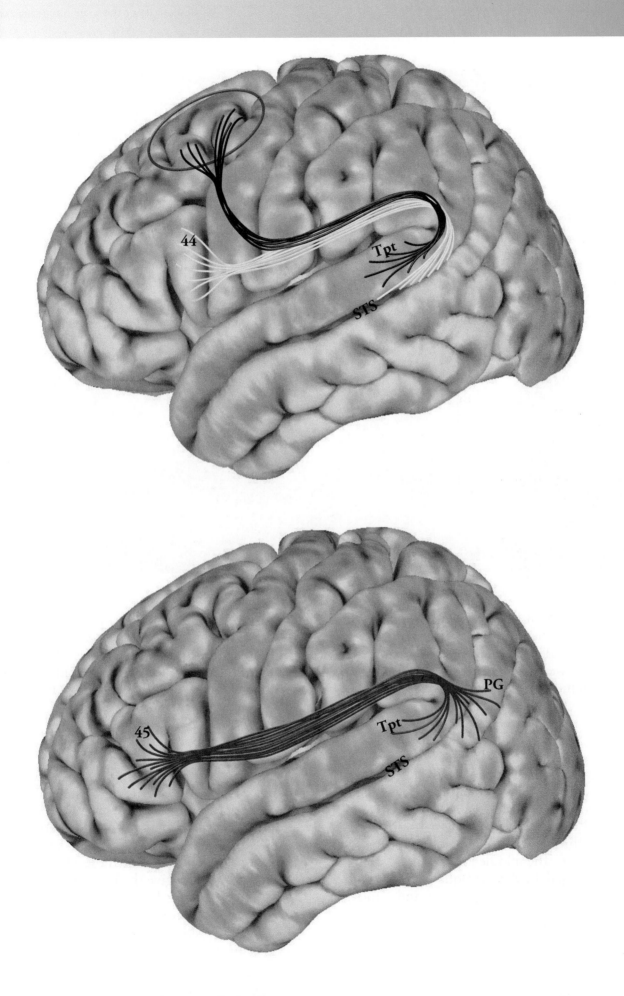

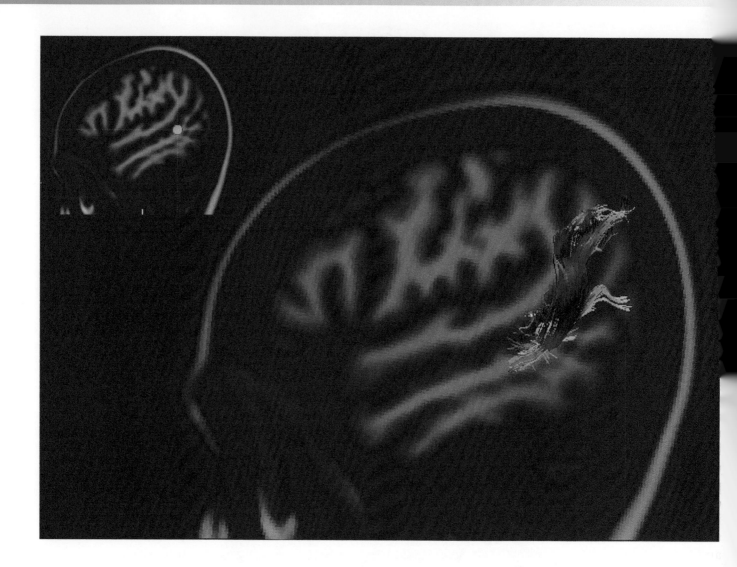

FIGURE 53 Diffusion MRI reconstruction in one brain of fibers coursing from the ventral part of the superior temporal sulcus to the inferior parietal cortex. These fibers are crossing close to the end of the lateral fissure and can easily be included in reconstructions of arcuate fasciculus fibers directed to the frontal lobe. The macaque monkey data, however, shows that these fibers terminate in the inferior parietal cortex. The placement of the seed is indicated on the sagittal section shown at the top left side of the figure. Diffusion MRI details, as in figure 38.

are coursing through the same region in the white matter of the inferior parietal lobule. The problem is compounded because a major system of temporo-parietal axons, the middle longitudinal fasciculus, runs through the same region at the end of the lateral fissure. The middle longitudinal fasciculus was discovered in the macaque monkey with anatomical tracer methods by Seltzer and Pandya (1984) and, more recently, has been reconstructed with diffusion MRI in the human brain (Makris et al., 2009). These axons have been shown to link various parts of the anterior and intermediate superior temporal gyrus and sulcus with the inferior parietal cortex and course through the white matter around the end of the lateral fissure (fibers in light blue in Figs. 50 and 51). Failure to separate these axons from those of the arcuate fasciculus proper can give the incorrect impression that fibers originating from anterior parts of the temporal lobe proceed posteriorly towards the end of the temporal lobe and then arch around the lateral fissure (arcuate fasciculus) to continue all ne way to the frontal lobe.

In conclusion, the white matter at the end of lateral fissure is a crossing point for many different axonal systems and it is not clear how many of the so-called arcuate fasciculus fibers are indeed monosynaptic axons originating within specific temporal cortical areas and terminating within particular frontal cortical areas and how many are in fact components of the middle longitudinal temporo-parietal system and the superior longitudinal parieto-frontal system. For the language researcher, the critical question is the following: Which of the many axon fascicles that are entering this crossroads of white matter actually terminate in Broca's region (areas 44 and 45)?

Petrides and Pandya (1988) examined the axonal pathways that originate from the superior temporal gyrus and are directed towards the frontal lobe in the macaque monkey using the autoradiographic method. Connections from the posterior third of the superior temporal gyrus arched around the lateral fissure but were directed towards the posterior dorsolateral frontal region and terminated primarily in area 8Ad and the adjacent rostral part of dorsal area 6 (area 6DR) (axons in dark blue in Fig. 52). There were no connections directed towards the ventrolateral frontal region (Petrides and Pandya, 1988). We interpreted these findings as the linkage between the auditory frontal attentional area 8Ad and a posterior auditory temporal area. At the time, we had not yet examined the cytoarchitecture of the ventrolateral frontal region of the macaque monkey in comparison with the human cortex and, therefore, we did not know whether the homologues of areas 44 and 45 existed in the macaque monkey brain. We therefore entertained the possibility of a species difference, although we were cognizant of the fact that good anatomical evidence linking the caudal third of the superior temporal gyrus with Broca's region in the human brain was lacking, despite the famous schematic diagram of Geschwind (Fig. 36).

In the 1990s, as part of our comparative anatomical examination of the human and the macaque frontal cortex, we were able to establish the existence of the cytoarchitectonic homologues of areas 44 and 45 in the macaque monkey (Petrides and Pandya, 1994, 2002; Petrides et al., 2005; Fig. 43). In a new series of studies to establish the connectivity of these areas in the macaque monkey with anatomical tracer injections, we were able to show that there were in fact three branches of the arcuate fasciculus originating from the posterior third of the superior temporal gyrus and adjacent superior temporal sulcal region (Fig. 52): 1) A branch originating from the dorsal part of the superior temporal gyrus close to the lateral fissure is directed to the posterior dorsolateral frontal cortex and terminates in area 8Ad and area 6DR (Petrides and Pandya, 1988, 2009; fibers in dark blue in Fig. 52). 2) A branch originating from the ventralmost part of the gyrus and the cortex in the upper bank of the superior temporal sulcus terminates in area 44 (Petrides and Pandya, 2009; Frey et al., 2013; fibers in yellow in Fig. 52). 3) A branch that originates from the caudal superior temporal gyrus and the adjacent superior temporal sulcus terminates in area 45 (Petrides and Pandya, 2009; Frey et al., 2013; fibers in dark red in Fig. 52). In other words, it was clear that we had earlier failed to demonstrate a projection from the posterior third of the superior temporal gyrus to the ventrolateral frontal cortex (Petrides and Pandya, 1988) because we had placed the injection dorsally on the gyrus close to the lateral fissure and not ventrally close to the superior temporal sulcus. Figure 52 provides our interpretation of the homologous connections in the human brain. These monosynaptic axonal connections originating from the posterior superior temporal region proceed to the frontal lobe as part of the arcuate fasciculus.

In conclusion, the macaque monkey data indicate that the arcuate fasciculus (if defined in the traditional manner as those monosynaptic axons that arch around the end of the lateral fissure to link temporo-parietal cortex with frontal cortex) comprises axons that originate only from the caudal part of the superior temporal gyrus and sulcus, i.e. primarily area Tpt and the ventral part of area PG (Fig. 52). However, one should be aware that reconstructions of the arcuate fasciculus with diffusion MRI can give the impression that it originates from more anterior parts of the temporal lobe because axons linking anterior temporal cortex with the inferior parietal lobule course close to the end of the lateral fissure in the region of the arcuate fasciculus (light blue fibers in Figs. 50 and 51). These temporo-parietal axons originate from anterior and intermediate parts of the superior temporal gyrus and sulcus (Seltzer and Pandya, 1984) and the cortex immediately below the superior temporal sulcus (e.g., Distler et al., 1993). Figure 53 provides diffusion MRI reconstruction of some of the temporo-parietal fibers that enter the inferior parietal region around the end of the lateral fissure.

Catani and colleagues (Catani et al., 2005; Catani and Thiebaut de Schotten, 2008) have dealt with the issue of axons that enter the arcuate fasciculus but terminate in the inferior parietal lobule and are not directed to the ventrolateral frontal region by making a distinction between a direct segment and an indirect segment of this fasciculus. One of the indirect segments is shown to connect parts of the lateral temporal region with the inferior parietal lobule and this may be the fibers of the middle longitudinal fasciculus directed to the inferior parietal lobule (Seltzer and Pandya, 1984; Makris et al., 2009) (Figs. 50 and 51, fibers in light blue). The other indirect component that leaves from the rostral inferior parietal lobule (Geschwind's territory) and is directed towards the ventrolateral frontal region may be the third branch of the superior longitudinal fasciculus (SLF III), demonstrated by Petrides and Pandya (1984) in the monkey and by Frey et al. (2008) and Makris et al. (2005) in DTI reconstructions in the human brain (Figs. 48 and 49, fibers in green).

The arcuate fasciculus would be expected to be less prominent in the macaque monkey for the following reasons. The cortex in the caudal third of the superior temporal sulcus and the adjacent caudal inferior parietal region from which axons are directed to the lateral frontal cortex is largely lying above the end of the lateral fissure because of the sharp ascending direction of the superior temporal sulcus in this primate species. Thus, most of the axons from the caudal parietal region and the caudal superior temporal sulcus will proceed directly as part of the superior longitudinal fasciculus (SLF II) with fewer axons having to arch around the end of the lateral fissure (arcuate fasciculus) in the monkey relative to the human in which the homologous region has expanded and shifted more ventrally. As is well known since the seminal studies of Shellshear (1927), the parieto-temporal junction region has undergone major

expansion in the human brain with large parts of the cortex lying in the depths of the caudal superior temporal sulcus in the monkey emerging on the surface, creating three branches of the caudal superior temporal sulcus, and shifting the location of homologous cortical areas ventral to the end point of the lateral fissure (see Petrides, 2012 and Segal and Petrides, 2012b). In correspondence with this increase in the cortical surface of the parieto-temporal junction, we would expect a major increase in the white matter volume in this region which is the crossroads of several pathways: superior longitudinal fasciculus II, arcuate fasciculus, middle longitudinal fasciculus, and superior occipito-frontal fasciculus. Diffusion MRI studies have confirmed the expansion of the arcuate fasciculus in the human in relation to the monkey (Glasser and Rilling, 2008; Rilling et al., 2008).

However, despite the enormous expansion of the parieto-temporal junction region in the human brain, resting state data indicate that the connectivity of this region remains fundamentally similar to that of the macaque monkey: it is the expanded PG region in the human brain (angular gyrus) that is co-activated with area 45 and, importantly, not the nearby areas PF and PFG (supramarginal gyrus), Tpt of superior temporal gyrus, or the posterior occipito-temporal cortex just ventral to the PG region (see Fig. 46; Margulies and Petrides, 2013). As is the case with the caudal parieto-temporal region, in the human brain, the anterior and intermediate multisensory cortical region of the superior temporal sulcus has expanded ventrally creating the new morphological entity known as the middle temporal gyrus. Monkey data with axonal tracers show that area 45 is strongly connected with the auditory related cortex of the ventral superior temporal gyrus (close to the sulcus), the multisensory cortex within the superior temporal sulcus, and the cortex immediately below the ventral lip of the superior temporal sulcus (Petrides and Pandya, 2009; Frey et al., 2013). Importantly, it is this auditory and expanded multisensory region of the human brain that is shown to be strongly connected with area 45 by resting state connectivity (Fig. 46; Margulies and Petrides, 2013). In other words, despite the major expansion of the parieto-temporal junction and the anterior-to-intermediate temporal region esulting in geographic shifts of homologous regions, the connectivity remains similar to that of the much less developed macaque monkey cortex. Note that the expansions in the temporo-parietal region of the human brain (e.g., area PG) are accompanied by corresponding expansions in the frontal lobe (e.g., area 45) and that these expanded regions maintain their fundamental connectivity between the macaque monkey and the human brains (e.g., Margulies and Petrides, 2009, 2013). However, there may be asymmetries in the strength of the connections, as suggested by Rilling and colleagues (2008).

In contrast to the differential connectivity patterns of the parieto-temporal and supero-lateral temporal regions with specific areas of the ventrolateral frontal region shown by resting state connectivity (consistent with the axonal tracing in monkeys), the diffusion MRI data indicates widespread diffusivity of the reconstructed human arcuate fasciculus into the entire ventrolateral frontal region (areas 6, 44, 45, and 47/12) (Rilling et al., 2008), suggesting a failure of the method to discriminate monosynaptic from polysynaptic connections. Similarly, the suggested connections of the ventrolateral frontal region via the arcuate fasciculus with anterior temporal areas may be failures of the method to discriminate many of the middle longitudinal fasciculus axons that mingle with the arcuate fasciculus to terminate in the inferior parietal lobule. Although possible, it appears counterintuitive that axons from anterior temporal cortex would course all the way to the parietal cortex only to turn around the end of the lateral fissure to continue to the ventrolateral frontal region, when they can simply proceed directly across the lateral fissure. These issues may be resolved in the future with the development of methods that can discriminate monosynaptic axons from polysynaptic links in the human brain. In this atlas, we present those connections via the arcuate fasciculus that are likely to be truly monosynaptic (Fig. 52) and also those additional temporo-parietal fibers mixed with the arcuate fasciculus that give rise to multistep links (polysynaptic) via other cortical areas (Fig. 51).

THE TEMPORO-FRONTAL EXTREME CAPSULE FASCICULUS: AN UNSUSPECTED TEMPORO-FRONTAL FASCICULUS FOR LANGUAGE PROCESSING

The temporo-frontal extreme capsule fasciculus was discovered by Petrides and Pandya (1988) in the macaque monkey in a study that examined, with the autoradiographic method, the long association pathways that link different parts of the superior temporal gyrus with the frontal lobe. The axons leaving the most anterior superior temporal gyrus were shown to course as part of the uncinate fasciculus and those from its most posterior part through the arcuate fasciculus. However, axons originating from the intermediate part of the superior temporal gyrus and the adjacent superior temporal sulcus were turning medio-dorsally to enter the extreme capsule and continue to the frontal lobe where they terminated within ventrolateral and dorsolateral frontal areas (Figs. 41 and 54). In other words, axons originating from the largest part of the superior temporal gyrus and sulcus were directed towards the lateral frontal cortex via the extreme capsule.

Although we reported these findings in 1988, they did not have a major impact on neuroanatomical investigations of language for two reasons. First, the neurolinguistic field was focused on the arcuate fasciculus as the major language pathway linking the temporal language zone (Wernicke's area) with the ventrolateral frontal region (Broca's region) and it was assumed that this newly reported pathway via the extreme capsule in the monkey had little relevance to language. Second, it was not known at the time that the ventrolateral frontal region of the macaque monkey includes areas that are cytoarchitectonically homologous to areas 44 and 45 (Broca's region) of the human brain. Indeed, we had used the old nomenclature of Walker (1940) to describe the ventrolateral frontal architectonic areas within which the axons of the newly discovered temporo-frontal extreme capsule fasciculus terminated. Soon after, however, in comparative studies of the human and the macaque monkey frontal cortex, we demonstrated the cytoarchitectonic homologues of areas 44 and 45 in the monkey (Petrides and Pandya, 1994, 2002; Petrides et al., 2005) (Fig. 43).

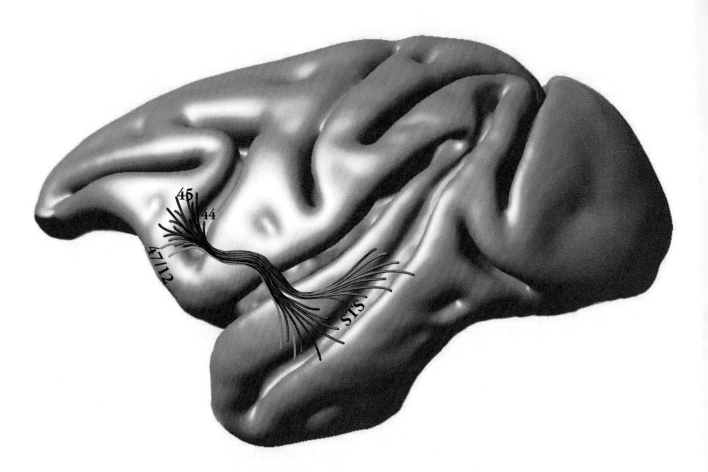

FIGURE 54 (ABOVE) Schematic illustration of the temporo-frontal extreme capsule fasciculus linking the intermediate region of the superolateral temporal lobe with the ventrolateral frontal area 45A. There are only minor connections with area 44 and 47/12. This fasciculus was first discovered in the macaque monkey (Petrides and Pandya, 1988, 2009), but there is now evidence for its existence in the human brain (Frey et al., 2008). **(RIGHT)** The temporo-frontal extreme capsule fasciculus projected onto the human average MRI brain.

Based on these findings, we decided to re-investigate the cortico-cortical association pathways that originate from the superior temporal and the inferior parietal region and are directed to the ventrolateral frontal region (Petrides and Pandya, 2009). This research revealed a complex network of connections, via distinct association fasciculi, from specific inferior parietal and superior temporal areas to the various areas of the ventrolateral frontal region, namely ventral premotor area 6, area 44, and area 45. In the ventrolateral frontal region, the major target of axons originating from an extensive zone along the intermediate part of the superior temporal gyrus and sulcus and coursing via the extreme capsule fasciculus was shown to be area 45, with only a few axons targeting areas 44 and 47/12 (Petrides and Pandya, 2009) (Fig. 54). Note that there are also connections between the intermediate superior temporal region and the dorsolateral frontal cortex, but these are not displayed in figure 54 which is focused on the ventrolateral frontal connections. For these additional dorsolateral connections, see figure 56 and articles by Petrides and Pandya (1988, 2007, 2009), Hackett et al. (1999), and Romanski et al. (1999).

In a diffusion tensor imaging study, we were able to provide evidence for the existence of the temporo-frontal extreme capsule fasciculus in the human brain (Frey et al., 2008) (Fig. 55). Other studies using diffusion MRI in the human brain also provided evidence for a ventral pathway that appears to run under the insula (e.g., Parker et al.,

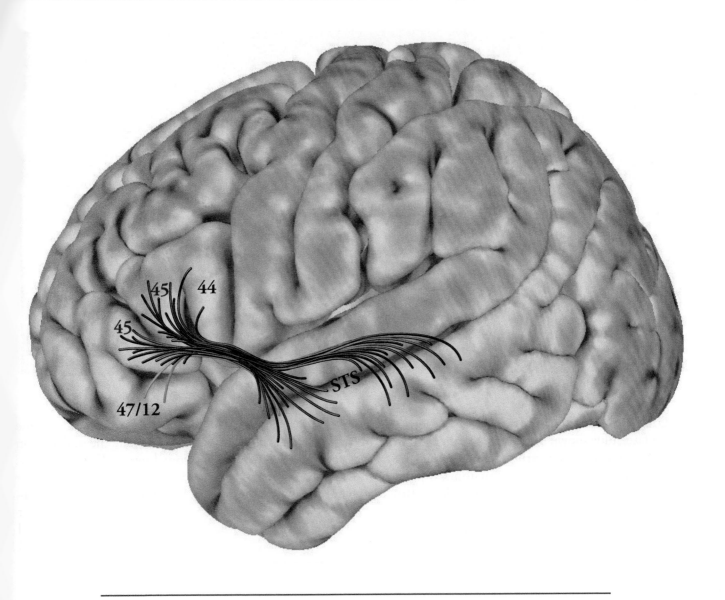

2005; Saur et al., 2008). In addition, the specific connectivity patterns of the various ventrolateral frontal areas with the inferior parietal lobule and the lateral temporal region has been explored with resting state connectivity (Kelly et al., 2010; Margulies and Petrides, 2013). This method establishes correlations of activity between different cortical areas and these correlations appear to reflect, at least partly, anatomical connectivity (e.g., Biswal et al., 1995). In a recent study, we were able to show that resting state activity of area 45 correlates with activity in the superior temporal sulcus and adjacent cortex of the superior and middle temporal gyri (Fig. 46). These cortical regions have been shown to be connected via the temporo-frontal extreme capsule fasciculus in the macaque monkey

(Petrides and Pandya, 1988, 2009).

A distinction is now frequently made between a dorsal and a ventral language stream (e.g., Hickok and Poeppel, 2004; Saur et al., 2008) by analogy to the well known dorsal and ventral visual (Ungerleider and Mishkin, 1982; Goodale and Milner, 1992) and auditory processing streams (Kaas and Hackett, 2000; Rauschecker and Tian, 2000; Poremba et al., 2003). The temporo-frontal extreme capsule fasciculus is now thought to be the major pathway of the ventral language stream, while the arcuate fasciculus/superior longitudinal fasciculus is thought to be the major pathway of the dorsal language stream (e.g., Saur et al., 2008).

Although it took considerable time to appreciate the relevance of the temporo-frontal extreme

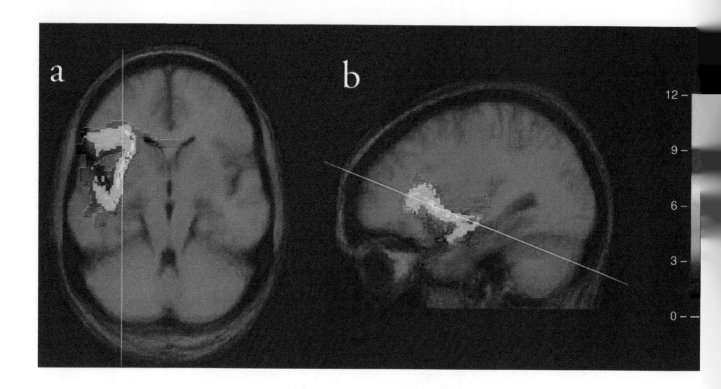

FIGURE 55 Group data showing diffusion tensor imaging (DTI) reconstruction of the temporo-frontal extreme capsule fasciculus shown on an axial (a) and a sagittal (b) MRI section. The regions of interest were the mid-section of the superior temporal gyrus and sulcus and Broca's region in the caudal inferior frontal gyrus. The projections are superimposed on the average anatomical MRI of the 12 subjects that constituted the data sample. The color intensity scales indicate overlap across subjects. Brighter colors signify greater overlap. (From Frey et al., 2008, with permission.)

capsule fasciculus to language studies, this is most probably the direct pathway that Wernicke (1881) thought to be the main connection between the temporal language comprehension zone and the ventrolateral frontal region. In figure 56, we present Wernicke's original schematic diagram with his speculation of a direct connection coursing under the insula and five cases in the macaque monkey in which we injected radioactively labeled amino acids in the temporal lobe to trace the axons to the frontal cortex (Petrides and Pandya, 2009). All these cases with injections in the lateral temporal region that extends from approximately the level of the rostral premotor cortex to the level of the anterior intraparietal sulcus (region marked by the two red lines in Wernicke's schematic diagram of the human brain) had labeled axons that coursed anteriorly via the extreme capsule fasciculus to

terminate in the ventrolateral frontal region. Thus, we conclude that the intermediate part of the supero-lateral temporal region is connected with the ventrolateral prefrontal region via the temporo-frontal extreme capsule fasciculus (Petrides and Pandya, 1988, 2009) (Figs. 54 and 56). The caudal part of the supero-lateral temporal cortical region is connected with ventrolateral frontal cortex via the arcuate fasciculus (see Fig. 52).

Why had classical gross dissection studies missed this major temporo-frontal fasciculus? We believe that these axons had not been noted before because they were overshadowed by the axonal system that was named the inferior occipito-frontal fasciculus by Curran (1909). As pointed out before, a major system of axons coursing from the occipital lobe across the temporal lobe towards the frontal lobe is one of the easiest fasciculi to demonstrate, both

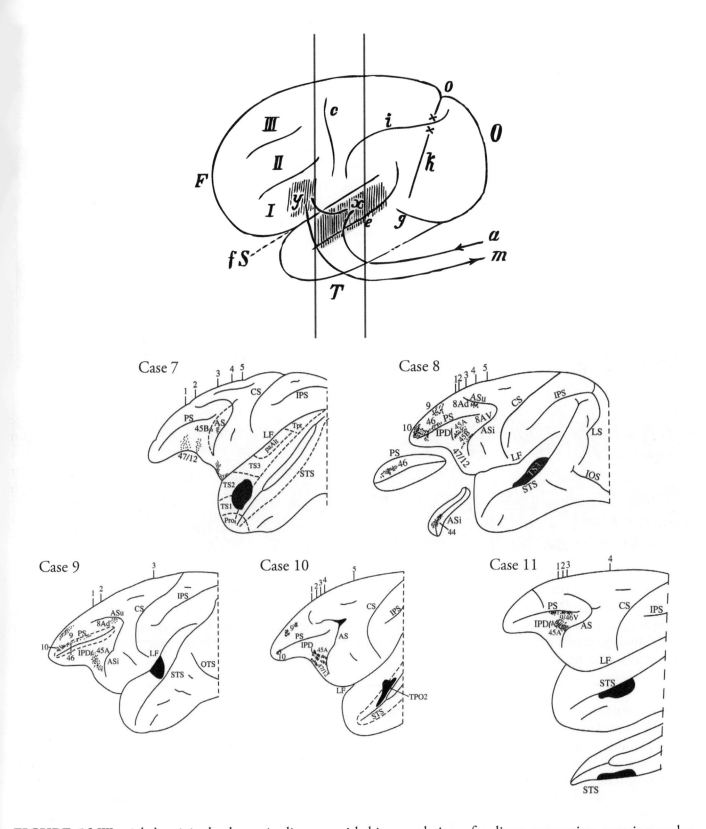

FIGURE 56 Wernicke's original schematic diagram with his speculation of a direct connection coursing under the insula and five cases in the macaque monkey with radioactively labeled amino acids injected into various parts of the intermediate superolateral temporal region. The corresponding region on Wernicke's schematic diagram is indicated in yellow between the two red vertical lines. In all five cases, axons could be traced from the injection site (shown in black) to the ventrolateral frontal cortex via the extreme capsule fasciculus (Petrides and Pandya, 2009).

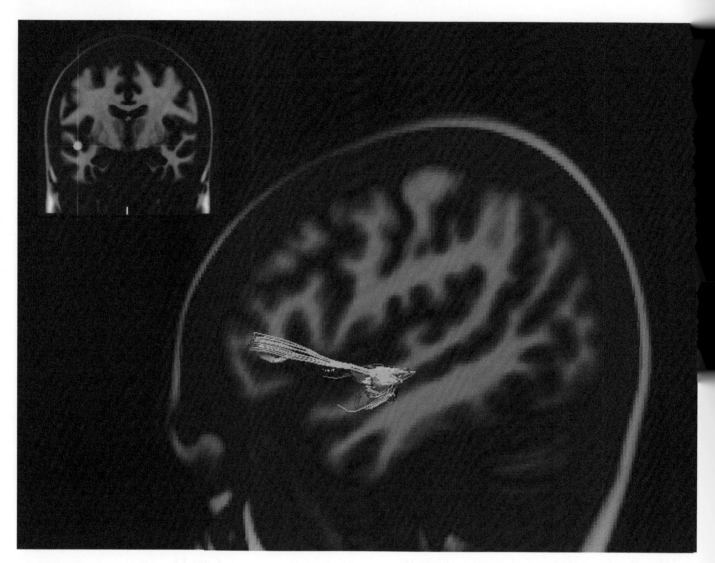

FIGURE 57 Reconstruction of the temporo-frontal extreme capsule fasciculus in one brain. Exclusion zones were set to prevent all contribution from the occipital and parietal regions lying posterior to the end of the lateral fissure and all contribution from the right hemisphere. A seed in the white matter of the intermediate part of the superior temporal gyrus (coronal section at the top left part of the image) resulted in the reconstruction of an axonal system leaving this region and directed via the ventral part of the extreme capsule towards the pars triangularis. Diffusion MRI details, as in figure 38.

with the gross dissection method and diffusion MRI (Figs. 37 and 38). The problem, however, is that it is not possible to know how many of the axons are truly monosynaptic occipito-frontal links and how many originate from lateral temporal cortex and, as they course towards the frontal cortex, mingle with those that are coming from the occipital region. Curran (1909), who first demonstrated the inferior occipito-frontal fasciculus, acknowledged the fact that many of the axons originate in the lateral temporal cortex. Figure 57 shows a reconstruction of the temporo-frontal extreme capsule fasciculus with diffusion MRI in one subject. In order to reconstruct this axonal system independent of the confounding contribution of occipital and parietal fibers, we placed two exclusion zones to prevent 1) all contribution of fibers from the occipital and parietal region lying posterior to the end of the lateral fissure and 2) all contribution from the right hemisphere. We then proceeded to place a seed in the white matter of the intermediate part of the superior temporal gyrus close to the region that we know from macaque monkey studies the temporo-frontal extreme capsule fasciculus courses.

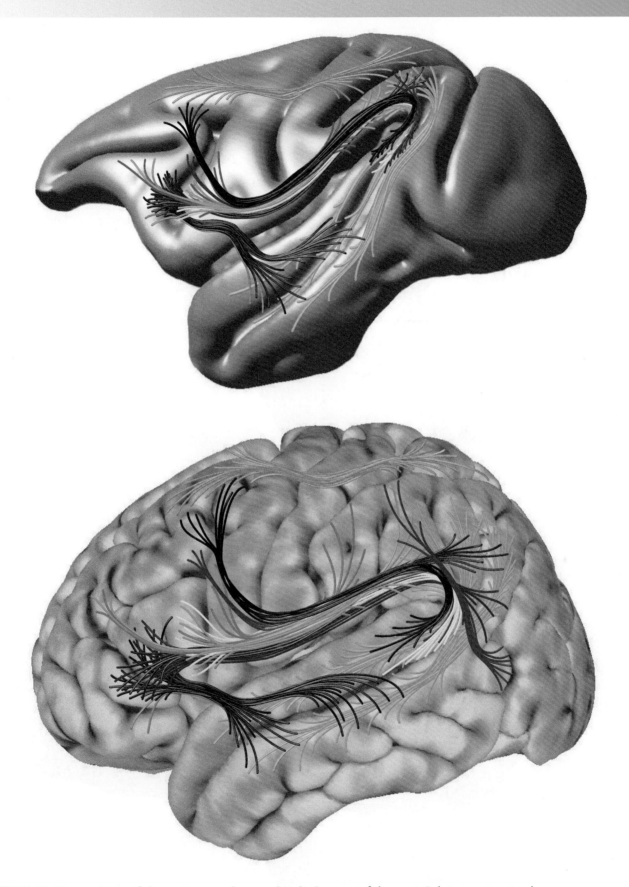

FIGURE 58 Illustrations of the various pathways that link areas of the peri-Sylvian region in the macaque monkey (above) and human brain (below). Color codes as in figures 48-52. An additional fascicle from occipito-temporal motion areas to the intraparietal sulcus and lateral frontal cortex is shown in scarlet red.

Area 45A

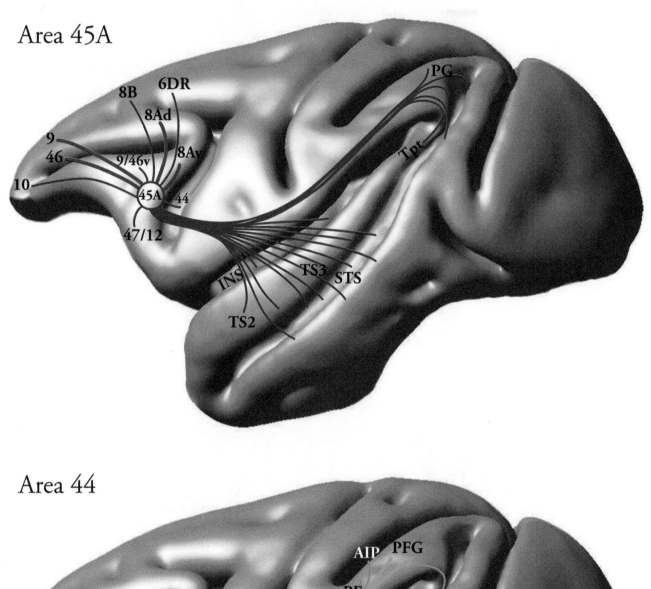

Area 44

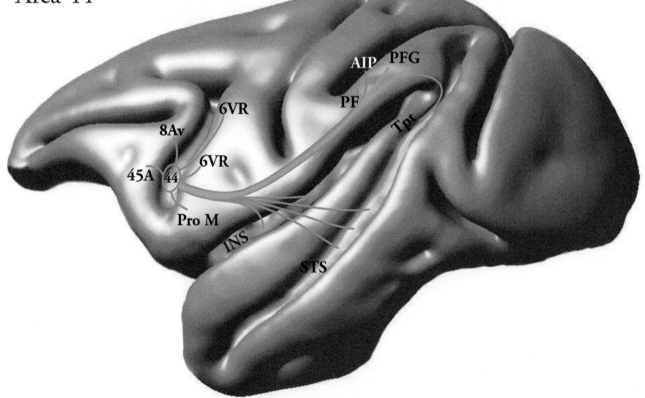

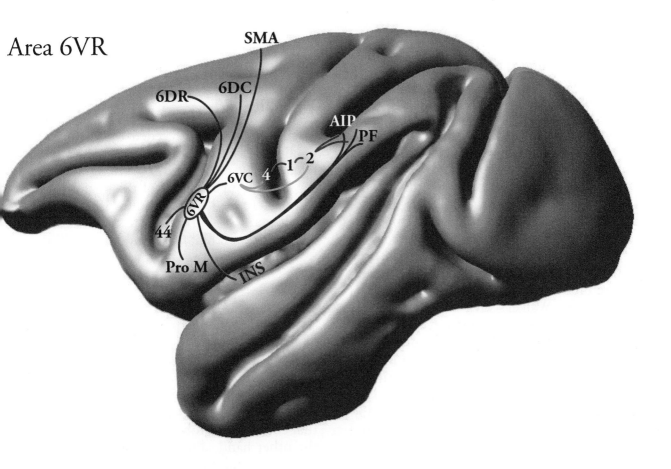

Area 6VR

FIGURE 59 (LEFT AND ABOVE) Schematic illustrations of the cortico-cortical axonal system centered on ventrolateral prefrontal area 45A, transitional frontal area 44, and rostroventral premotor area 6VR. Abbreviations: INS, insula; STS, superior temporal sulcus.

---

n axonal system leaving this region and directed a the ventral part of the extreme capsule towards e pars triangularis was thus reconstructed (Fig. 7). Since we have excluded any fibers from the ght hemisphere and the occipital and parietal gions within the left hemisphere, we can assume at a part of the temporo-frontal extreme capsule sciculus has been reconstructed.

CONCLUDING COMMENTS

The macaque monkey studies with anatomi- l tracers have provided important information constrain interpretations of the cortico-cor- cal axonal fibers reconstructed with diffusion 1RI. They have helped highlight the probable

monosynaptic cortico-cortical links between language relevant areas via particular fasciculi. Figure 58 presents our current understanding of the monosynaptic cortico-cortical links made between various areas of the peri-Sylvian region in the macaque monkey and our interpretation of these links in the human brain. We now turn to a brief presentation of the essential features of the connectivity of the two areas that comprise Broca's region: area 44 and area 45A.

The establishment of the homologues of these areas in the macaque monkey (Petrides and Pandya, 1994, 2002; Petrides et al., 2005) has permitted the examination of the precise connectivity patterns of these two areas. We believe that the cytoarchitectonic homologue of area 44 had not been identified before in the macaque monkey because it is hidden in the fundus of the lower limb of the arcuate sulcus (Fig. 43) and the usual coronal sectioning of this region distorts its cytoarchitectonic features. Indeed, we were able to identify this area reliably only when we started cutting the cortex perpendicular to the direction of the sulcus and thus obtained sections that provided a clear view of the cytoarchitectonic landscape of this region. The definition of area 45 in the macaque monkey generated considerable confusion. The macaque monkey cortex that corresponds cytoarchitectonically to area 45 of the human brain had been variously assigned, in older studies, to different areas of the prefrontal cortex (e.g., ventral part of Walker's area 46, Walker's area 12) (see Petrides and Pandya, 2002). Our comparative cytoarchitectonic studies identified the homologous area in the monkey by applying the same criteria in both species. In the monkey, it is located anterior to area 44 and extends onto the ventrolateral frontal region as far as the infraprincipal dimple (Fig. 43). Examination of the connections of areas 44 and 45A in the macaque monkey revealed a number of interesting facts (Fig. 59).

AREA 45A: THE GREAT PREFRONTAL INTEGRATOR

Area 45A is strongly connected with 1) the intermediate supero-lateral temporal region via the temporo-frontal extreme capsule fasciculus, 2) the caudal superior temporal region via the arcuate fasciculus, and 3) the caudal inferior parietal lobule (area PG) via the second branch of the superior longitudinal fasciculus (Fig. 59). Within the frontal cortex, area 45A has strong connections with ventrolateral, dorsolateral, and orbital frontal areas. No other area of the prefrontal cortex exhibits this widespread connectivity. Thus, this cytoarchitectonically unique area has access to auditory and multisensory information via its links to the supero-lateral temporal region which, in the left hemisphere of the human brain, subserves semantic processing (Binder et al., 2009). We have argued that this area may be the epicenter of ventrolateral prefrontal active retrieval mechanisms (Petrides, 2002). Area 45A is in a position to retrieve selectively information available within auditory, visual

and multisensory lateral temporal cortex and suc retrieval may be the essential feature that make this cortical area indispensable to language pro cessing in the left hemisphere (Petrides, 2006). Al discourse must be preceded by selective retrieva of the information to be articulated. There is nov considerable evidence from functional neuroimag ing for the involvement of this region in selectiv retrieval (e.g., Petrides et al., 1995; Thompson-Schill et al., 1997; Bokde et al., 2001; Kostopoulo and Petrides, 2003; Grindrod et al., 2008).

AREA 44: THE INTERMEDIARY BETWEEN COGNITIVE RETRIEVAL AND ARTICULATION

Despite the widespread connectivity that area 45A exhibits within the frontal cortex, it lacks direct connections to the ventral precentral gyrus and, therefore, direct access to the motor systems that control the orofacial musculature. Area 44, on the other hand, does have connections with both area 45A and the ventral premotor system and may serve as a transitional area between the pure motor system and the cognitive prefrontal retrieval system, acting as a go-between for the translation of retrieved information into action (Petrides, 2006). As can be seen in figure 59, premotor area 6VR, with which area 44 is connected, has widespread connections with various motor structures (Luppino et al., 1999; Gerbella et al., 2011). Area 6VR is also known as area F5 (Belmalih et al., 2009) and is the ventrolateral premotor area where the mirror neurons were initially discovered (Rizzolatti and Luppino, 2001). Thus, there is a gradation of areas in the ventrolateral frontal region, ranging from the most cognitive area 45A to the action coordination area 6VR via the intermediary of area 44. It is therefore of interest to note that area 44 and area PFG (in the supramarginal gyrus), which are strongly interconnected, have been considered to be critical parts of the phonological system (Paulesu et al., 1993; Church et al., 2011).

References

Ad-Dab'bagh, Y., Lyttelton, O., Muehlboeck, J.S., Lepage, C., Einarson, D., Mok, K., Ivanov, O., Vincent, R.D., Lerch, J., Fombonne, E., and Evans, A.C. (2006). The CIVET image processing environment: a fully automated comprehensive pipeline for anatomical neuroimaging research. In Proceedings of the 12th Annual Meeting of the Organization for Human Brain Mapping (pp. S45). Florence, Italy.

Axer, M., Amunts, K., Grässel, D., Palm, C., Dammers, J., Axer, H., Pietrzyk, U., and Zilles, K. (2011). A novel approach to the human connectome: Ultra-high resolution mapping of fiber tracts in the brain. NeuroImage, 54, 1091-1101.

Ackermann, H. and Riecker, A. (2004). The contribution of the insula to motor aspects of speech production: A review and a hypothesis. Brain and Language, 89, 320-328.

Amiez, C. and Petrides, M. (2012). Neuroimaging evidence of the anatomo-functional organization of the human cingulate motor areas. Cerebral Cortex, Nov 6. [Epub ahead of print]

Amunts, K., Schleicher, A., Burgel, U., Mohlberg, H., Uylings, H.B.M., and Zilles, K. (1999). Broca's region revisited: cytoarchitecture and intersubject variability. Journal of Comparative Neurology, 412, 319-341.

Amunts, K., Lenzen, M., Friederici, A.D., Schleicher, A., Morosan, P., Palomero-Gallagher, N., and Zilles, K. (2010). Broca's region: novel organizational principles and multiple receptor mapping. Public Library of Science Biology, 8, e1000489.

Badre, D. and Wagner, A.D. (2007). Left ventrolateral prefrontal cortex and the cognitive control of memory. Neuropsychologia, 45, 2883-2901.

Baldo, J.V., Wilkins, D.P., Ogar, J., Willock, S., and Dronkers, N.F. (2011). Role of the precentral gyrus of the insula in complex articulation. Cortex, 47, 800-807.

Basso, A., Lecours, A.R., Moraschini, S., and Vanier, M. (1985). Anatomoclinical correlations of the aphasias as defined through computerized tomography: Exceptions. Brain and Language, 26, 201-229.

Bates, E., Friederici, A., and Wulfeck, B. (1987). Grammatical morphology in aphasia: evidence from three languages. Cortex, 23, 545-574.

Belmalih, A., Borra, E., Contini, M., Gerbella, M., Rozzi, S., and Luppino, G. (2009). Multimodal architectonic subdivision of the rostral part (area F5) of the macaque ventral premotor cortex. Journal of Comparative Neurology, 512, 183-217.

Bernal, B. and Altman, N. (2010). The connectivity of the superior longitudinal fasciculus: a tractography DTI study. Magnetic Resonance Imaging, 28, 217-225.

Betz, W. (1874). Anatomischer Nachweis zweier Gehirncentra. Centralblatt für die medicinischen Wissenschaften, 12, 578-580 (see also pp. 595-599).

Binder, J.R., Frost, J.A., Hammeke, T.A., Cox, R.W., Rao, S.M., and Prieto, T. (1997). Human brain language areas identified by functional magnetic resonance imaging. Journal of Neuroscience, 17, 353-362.

Binder, J.R., Desai, R.H., Graves, W.W., and Conant, L.L. (2009). Where is the semantic system? A critical review and meta-analysis of 120 functional neuroimaging studies. Cerebral Cortex, 19, 2767-2796.

Biswal, B., Yetkin, F.Z., Haughton, V.M., and Hyde, J.S. (1995). Functional connectivity in the motor cortex of resting human brain using echo-planar MRI. Magnetic Resonance in Medicine, 34, 537-541.

Bokde, A.L.W., Tagamets, M-A., Friedman, R.B., and Horwitz, B. (2001). Functional interactions of the inferior frontal cortex during the processing of words and word-like stimuli. Neuron, 30, 609-617.

Bogen, J.E. and Bogen, G.M. (1976). Wernicke's region – Where is it? Annals of the New York Academy of Sciences, 280, 834-843.

Borovsky, A., Saygin, A.P., Bates, E., and Dronkers, N. (2007). Lesion correlates of conversational speech production deficits. Neuropsychologia, 45, 2525-2533.

Broca, P. (1861a). Perte de la parole, ramollissement chronique et destruction partielle du lobe antérieur gauche du cerveau. Bulletins de la Société d'Anthropologie de Paris, 2, 235-238.

Broca, P. (1861b). Remarques sur le siège de la faculté du langage articulé, suivies d'une observation d'aphémie (perte de la parole). Bulletins de la Société Anatomique de Paris, 6, 330-357.

Broca, P. (1861c). Nouvelle observation d'aphémie produite par une lésion de la moitié postérieure des deuxième et troisième circonvolutions frontales. Bulletins de la Société Anatomique de Paris, 6, 398-407.

Brodmann, K. (1908). Beitraege zur histologischen Lokalisation der Grosshirnrinde. VI. Mitteilung: Die Cortexgliederung des Menschen. Journal für Psychologie und Neurologie (Leipzig), 10, 231-246.

Brodmann, K. (1909). Vergleichende Lokalisationslehre der Grosshirnrinde in ihren Prinzipien dargestellt auf Grund des Zellenbaues. Leipzig: Barth.

Cabanis, E.A., Iba-Zizen, M.T., Abelanet, R., Monod-Broca, P., and Signoret, J.L. (1994). "Tan-Tan" the first Paul Broca's patient with

"Aphemia" (1861): CT (1979) and MRI (199. of the brain. In L. Picard and G. Salamon (Eds. 4th Refresher Course of the ESNR: Languag and Aphasias (pp. 9-22). European Society c Neuroradiology, Nancy, France.

Cabeza, R. and Kingstone, A. (Eds.). (2006: Handbook of Functional Neuroimaging o Cognition. Cambridge, MA: MIT Press.

Campbell, A. W. (1905). Histological Studie on the Localisation of Cerebral Function Cambridge: Cambridge University Press.

Campbell, J.S.W. and Pike, G.B. (2013). Potentia and limitations of diffusion MRI tractography for the study of language. Brain and Language. in press.

Caplan, D. (1992). Language: Structure, Processing, and Disorders. Cambridge, MA: MIT Press.

Caplan, D. and Hildebrandt, H. (1998). Disorders of Syntactic Comprehension. Cambridge, MA: MIT Press.

Caspers, S., Geyer, S., Schleicher, A., Mohlberg, H., Amunts, K., and Zilles, K. (2006). The human inferior parietal cortex: cytoarchitectonic parcellation and interindividual variability. Neuroimage, 33, 430-448.

Caspers, S., Eickhoff, S.B., Geyer, S., Scheperjans, F., Mohlberg, H., Zilles, K., and Amunts, K. (2008). The human inferior parietal lobule in stereotaxic space. Brain Structure and Function, 212, 481-495.

Castaigne, P., Lhermitte, F., Signoret, J.L., and Abelanet, R. (1980). Description et étude scanographique du cerveau de Leborgne: la découverte de Broca. Revue Neurologique (Paris), 136, 563-583.

Catani, M. and Mesulam, M. (2008). The arcuate fasciculus and the disconnection theme in language and aphasia: history and current state. Cortex, 44, 953-961.

Catani, M. and Thiebaut de Schotten, M. (2008). A diffusion tensor imaging tractography atlas for virtual in vivo dissections. Cortex, 44, 1105-1132.

Catani, M., Jones, D.K., and ffytche, D.H. (2005). Perisylvian language networks of the human brain. Annals of Neurology, 57, 8-16.

Chapados, C. and Petrides, M. (2013). Impairment only on the fluency subtest of the Frontal Assessment Battery after prefrontal lesions. Brain, in press.

Church, J.A., Balota, D.A., Petersen, S.E., and Schlaggar, B.L. (2011). Manipulation of length and lexicality localizes the functional neuro-anatomy of phonological processing in adult readers. Journal of Cognitive Neuroscience, 23, 1475-1493.

Cohen, L. and Dehaene, S. (2004). Specialization within the ventral stream: the case for the visual word form area. NeuroImage, 22, 466-476.

Cohen, L., Dehaene, S., Naccache, L., Lehéricy, S., Dehaene-Lambertz, G., Hénaff, M-A., and Michel, F. (2000). The visual word form area. Spatial and temporal characterization of an initial stage of reading in normal subjects and posterior split-brain patients. Brain, 123, 291-307.

Collins, D.L., Neelin, P., Peters, T.M., and Evans, A.C. (1994). Automatic 3D inter-subject registration of MR volumetric data in standardized Talairach space. Journal of Computer Assisted Tomography, 18, 192-205.

Collins, D.L. (2012). Montreal Neurological Institute (MNI) space. In M. Petrides, The Human Cerebral Cortex. An MRI Atlas of the Sulci and Gyri in MNI Stereotaxic Space (pp. 12-17). New York: Academic Press.

Colon-Perez, L.M., Spindler, C., Goicochea, S., Triplett, W., Parekh, M., Montie, E., Carney, P.R., and Mareci, T. (2012). Brain network metric derived from DWI: application to the limbic system. In Proceedings ISMRM 20th Scientific Meeting (pp. 651). Melbourne.

Cowan, W.M., Gottlieb, D.I., Hendrickson, A.E., Price, J.L., and Woolsey, T.A. (1972). The autoradiographic demonstration of axonal connections in the central nervous system. Brain Research, 37, 21-51.

Croxson, P.L., Johansen-Berg, H., Behrens, T.E., Robson, M.D., Pinsk, M.A., Gross, C.G., Richter, W., Richter, M.C., Kastner, S., and Rushworth, M.F. (2005). Quantitative investigation of connections of the prefrontal cortex in the human and macaque using probabilistic diffusion tractography. Journal of Neuroscience, 25, 8854-8866.

Curran, E.J. (1909). A new association fiber tract in the cerebrum with remarks on the fiber tract dissection method of studying the brain. Journal of Comparative Neurology and Psychology, 19, 645-656.

Dejerine, J. (1891a). Contribution à l'étude des troubles de l'écriture chez les aphasiques. Comptes Rendus Hebdomadaires des Séances et Mémoires de la Société de Biologie, 43, 97-113.

Dejerine, J. (1891b). Sur un cas de cécité verbale avec agraphie, suivi d'autopsie. Comptes Rendus Hebdomadaires des Séances et Mémoires de la Société de Biologie, 43, 197-201.

Dejerine, J. (1892). Contribution à l'étude anatomo-pathologique et clinique des différentes variétés de cécité verbale. Comptes Rendus Hebdomadaires des Séances et Mémoires de la Société de Biologie, 44, 61-90.

Dejerine, J. (1895). Anatomie des Centres Nerveux. Paris: Rueff et Cie.

Démonet, J-F., Thierry, G., and Cardebat, D. (2005). Renewal of the neurophysiology of language: Functional neuroimaging. Physiological Reviews, 85, 49-95.

Denes, G. (2011). Talking Heads: The

Neuroscience of Language. (P.V. Smith, Trans.). Sussex Psychology Press. (Original work published 2009).

De Ribaupierre, S., Fohlen, M., Bulteau, C., Dorfmüller, G., Delalande, O., Dulac, O., Chiron, C., and Hertz-Pannier, L. (2012). Presurgical language mapping in children with epilepsy: Clinical usefulness of functional magnetic resonance imaging for the planning of cortical stimulation. Epilepsia, 53, 67-78.

DeWitt, I. and Rauschecker, J.P. (2012). Phoneme and word recognition in the auditory ventral stream. Proceedings of the National Academy of Sciences USA, 109, E505-E514.

Distler, C., Boussaoud, D, Desimone, R., and Ungerleider, L.G. (1993). Cortical connections of inferior temporal area TEO in macaque monkeys. Journal of Comparative Neurology, 334, 125-150.

Dronkers, N.F. (1996). A new brain region for coordinating speech articulation. Nature, 384, 159-161.

Dronkers, N.F., Redfern, B.B., and Ludy, C.A. (1995). Lesion localization in chronic Wernicke's aphasia. Brain and Language, 51, 62-65.

Dronkers, N.F., Plaisant, O., Iba-Zizen, M.T., and Cabanis, E.A. (2007). Paul Broca's historic cases: high resolution MR imaging of the brains of Leborgne and Lelong. Brain, 130, 1432-1441.

Dronkers, N.F., Wilkins, D.P., Van Valin, R.D.Jr., Redfern, B.B., and Jaeger, J.J. (2004). Lesion analysis of the brain areas involved in language comprehension. Cognition, 92, 145-177.

Duffau, H. (2007). Contribution of cortical and subcortical electrostimulation in brain glioma surgery: methodological and functional considerations. Neurophysiologie Clinique, 37, 373-382.

Duffau, H. (2008). The anatomo-functional connectivity of language revisited: New insights provided by electrostimulation and tractograp phy. Neuropsychologia, 46, 927-934.

Duffau, H., Gatignol, P., Mandonnet, E., Peruzzi P., Tzourio-Mazoyer, N., and Capelle, L. (2005). New insights into the anatomo-functional con nectivity of the semantic system: a study using cortico-subcortical electrostimulations. Brain 128, 797-810.

Duffau, H., Gatignol, P., Mandonnet, E., Capelle L., and Taillandier, L. (2008). Intraoperative subcortical stimulation mapping of language pathways in a consecutive series of 115 patients with grade II glioma in the left dominant hemisphere. Journal of Neurosurgery, 109, 461-471.

Duffau, H., Gatignol, P., Moritz-Gasser, S., and Mandonnet, E. (2009). Is the left uncinate fasciculus essential for language? A cerebral stimulation study. Journal of Neurology, 256, 382-389.

Economo, C. and Koskinas, G.N. (1925). Die Cytoarchitektonik der Hirnrinde des erwachsenen Menschen. Wien und Berlin: Springer.

Exner, S. (1881). Untersuchungen über die Localisation der Functionen in der Grosshirnrinde des Menschen. Vienna: W. Braumuller.

Faust, M. (Ed.). (2012). The Handbook of the Neuropsychology of Language. Malden, MA: Blackwell Publishing.

Fink, R.P. and Heimer, L. (1967). Two methods for selective silver impregnation of degenerating axons and their synaptic endings in the central nervous system. Brain Research, 4, 369-374.

Frey, S., Mackey, S., and Petrides, M. (2013). Cortico-cortical connections of areas 44 and 45B in the macaque monkey. Brain and Language, in press.

Frey, S., Campbell, J.S., Pike, G.B., and Petrides, M. (2008). Dissociating the human language pathways with high angular resolution diffusion fiber tractography. Journal of Neuroscience, 28,11435-11444.

ried, I., Katz, A. McCarthy, G., Sass, K.J., Williamson, P., Spencer, S.S., and Spencer, D.D. (1991). Functional organization of human supplementary motor cortex studied by electrical stimulation. Journal of Neuroscience, 11, 3656-3666.

Friederici, A.D. (2009). Pathways to language: fiber tracts in the human brain. Trends in Cognitive Sciences, 13, 175-181.

Friederici, A.D. (2011). The brain basis of language processing: from structure to function. Physiological Reviews, 91, 1357-1392.

Friederici, A.D., Bahlmann, J., Heim, S., Schubotz, R.I., and Anwander, A. (2006). The brain differentiates human and non-human grammars: functional localization and structural connectivity. Proceedings of the National Academy of Sciences USA, 103, 2458-2463.

Fullerton, B.C. and Pandya, D.N. (2007). Architectonic analysis of the auditory-related areas of the superior temporal region in human brain. Journal of Comparative Neurology, 504, 470-498.

Gerbella, M., Belmalih, A., Borra, E., Rozzi, S., and Luppino, G. (2011). Cortical connections of the anterior (F5a) subdivision of the macaque ventral premotor area F5. Brain Structure and Function, 216, 43-65.

Germann, J., Robbins, S., Halsband, U., and Petrides, M. (2005). Precentral sulcal complex of the human brain: Morphology and statistical probability maps. Journal of Comparative Neurology, 493, 334-356.

Geschwind, N. (1970). The organization of language and the brain. Science, 170, 940-944.

Glasser, M.F. and Rilling, J.K. (2008). DTI tractography of the human brain's language pathways. Cerebral Cortex, 18, 2471-2482.

Goldberg, G. (1985). Supplementary motor area structure and function: Review and hypotheses. Behavioral Brain Sciences, 8, 567-588.

Goodale, M.A. and Milner, A.D. (1992). Separate visual pathways for perception and action. Trends in Neurosciences, 15, 20-25.

Goodglass, H. (1993). Understanding Aphasia. San Diego: Academic Press.

Grabner, G., Janke, A.L., Budge, M.M., Smith, D., Pruessner, J., and Collins, D.L. (2006). Symmetric atlasing and model based segmentation: an application to the hippocampus in older adults. In R. Larsen, M. Nielsen, and J. Sporring (Eds.), Medical Image Computing and Computer-Assisted Intervention – MICCAI 2006 (Vol. 4191, pp. 58-66). Berlin: Springer.

Grindrod, C.M., Bilenko, N.Y., Myers, B.E., and Blumstein, S.E. (2008). The role of the inferior frontal gyrus in implicit semantic competition and selection: An event related fMRI study. Brain Research, 1229, 167-178.

Grodzinsky, Y. (2000). The neurology of syntax: Language use without Broca's area. Behavioral and Brain Sciences, 23, 1-71.

Gullberg, M. and Indefrey, P. (Eds.). (2006). The Cognitive Neuroscience of Second Language Acquisition. Oxford: Blackwell Publishing.

Hackett, T.A., Stepniewska, I., and Kaas, J.H. (1999). Prefrontal connections of the parabelt auditory cortex in macaque monkeys. Brain Research, 817, 45-58.

Hagmann, P., Cammoun, L., Gigandet, X., Meuli, R., Honey, C.J., Wedeen, V.J., and Sporns, O. (2008). Mapping the structural core of human cerebral cortex. Public Library of Science Biology, 6, 1479-1493.

Hickok, G. and Poeppel, D. (2004). Dorsal and ventral streams: a framework for understanding aspects of the functional anatomy of language. Cognition, 92, 67-99.

Hillis, A.E. (Ed.). (2002). Handbook of Adult Language Disorders: Integrating Cognitive Neuropsychology, Neurology, and Rehabilitation. New York: Psychology Press.

Ingram, J.C.L. (2007). Neurolinguistics: An Introduction to Spoken Language Processing and its Disorders. New York: Cambridge University Press. .

Jackendoff, R. (2002). Foundations of Language: Brain, Meaning, Grammar, Evolution. New York: Oxford University Press.

Jbabdi, S. and Johansen-Berg, H. (2011). Tractography: Where do we go from here? Brain Connectivity, 1, 169-183.

Johansen-Berg, H. and Behrens, T.E. (2006). Just pretty pictures? What diffusion tractography can add in clinical neuroscience. Current Opinion in Neurology, 19, 379-385.

Jones, D.K. (2010). Challenges and limitations of quantifying brain connectivity in vivo with diffusion MRI. Imaging in Medicine, 2, 341-355.

Jones, D.K. and Cercignani, M. (2010). Twenty-five pitfalls in the analysis of diffusion MRI data. NMR in Biomedicine, 23, 803-820.

Jones, D.K., Knösche, T.R., and Turner, R. (2013). White matter integrity, fiber count, and other fallacies: The do's and don'ts of diffusion MRI. NeuroImage, 73, 239-254.

Kaas, J.H. and Hackett, T.A. (2000). Subdivisions of auditory cortex and processing streams in primates. Proceedings of the National Academy of Sciences USA, 97, 11793-11799.

Kaplan, E., Naeser, M.A., Martin, P.I., Ho, M., Wang, Y., Baker, E., and Pascual-Leone, A. (2010). Horizontal portion of arcuate fasciculus fibers track to pars opercularis, not pars triangularis, in right and left hemispheres: a DTI study. NeuroImage, 52, 436-444.

Kelly, C., Uddin, L.Q., Shehzad, Z., Margulies, D.S., Castellanos, F.X., Milham, M.P., an Petrides, M. (2010). Broca's region: linkin human brain functional connectivity data an non-human primate tracing anatomy studie European Journal of Neuroscience, 32, 383-398

Klingler, J. (1935). Erleichterung der makros kopischen Präparation des Gehirns durch de Gefrierprozeß. Schweizer Archiv für Neurologi und Psychiatrie, 36, 247-256.

Klingler, J. and Gloor, P. (1960). The connec tions of the amygdala and of the anterior tem poral cortex in the human brain. Journal o Comparative Neurology, 115, 333-369.

Kostopoulos, P. and Petrides, M. (2003). The mid ventrolateral prefrontal cortex: insights into its role in memory retrieval. European Journal of Neuroscience, 17, 1489-1497.

Krainik, A., Lehéricy, S., Duffau, H., Capelle, L., Chainay, H., Cornu, P., Cohen, L., Boch, A.L., Mangin, J.F., Le Bihan, D., and Marsault, C. (2003). Post-operative speech disorder after medial frontal surgery: role of the supplementary motor area. Neurology, 60, 587-594.

Kuypers, H.G.J.M. and Huisman, A.M. (1984). Fluorescent neuronal tracers. In S. Ferdoroff (Ed.), Advances in Cellular Neurobiology (Vol. 5, pp. 307-340). London: Academic Press.

Lewis, B.W. and Clarke, H. (1878). The cortical lamination of the motor area of the brain. Proceedings of the Royal Society, London, 27, 38-49.

Lichtheim, L. (1885). On aphasia. Brain, 7, 433-484.

Ludwig, E. and Klingler, J. (1956). Atlas Cerebri Humani. Basel: Karger.

Luppino, G., Murata, A., Govoni, P., and Matelli, M. (1999). Largely segregated parietofrontal connections linking rostral intraparietal cortex (areas AIP and VIP) and the ventral premotor cortex (areas F5 and F4). Experimental Brain

Research, 128, 181-187.

Makris, N., Kennedy, D.N., McInerney, S., Sorensen, A.G., Wang, R., Caviness, V.S.Jr., and Pandya, D.N. (2005). Segmentation of subcomponents within the superior longitudinal fascicle in humans: a quantitative, in vivo, DT-MRI study. Cerebral Cortex, 15, 854-869.

Makris, N., Papadimitriou, G.M., Kaiser, J.R., Sorg, S., Kennedy, D.N., and Pandya, D.N. (2009). Delineation of the middle longitudinal fascicle in humans: a quantitative, in vivo, DT-MRI study. Cerebral Cortex, 19, 777-785.

Margulies, D.S. and Petrides, M. (2013). Distinct parietal and temporal connectivity profiles of ventrolateral frontal areas involved in language production. Journal of Neuroscience, in press.

Margulies, D.S., Vincent, J.L., Kelly, C., Lohmann, G., Uddin, L.Q., Biswal, B.B., Villringer, A., Castellanos, F.X., Milham, M.P., and Petrides, M. (2009). Precuneus shares intrinsic functional architecture in humans and monkeys. Proceedings of the National Academy of Sciences USA, 106, 20069-20074.

Marie, P. (1906). Révision de la question de l'aphasie: La troisième circonvolution frontale gauche ne joue aucun rôle spécial dans la fonction du langage. Semaine Médicale, 26, 241-247.

Marie, P. and Foix, C. (1917). Les aphasies de guerre. Revue Neurologique, 24, 53-87.

Martin, R.E., MacIntosh, B.J., Smith, R.C., Barr, A.M., Stevens, T.K., Gati, J.S., and Menon, R.S. (2004). Cerebral areas processing swallowing and tongue movement are overlapping but distinct: a functional magnetic resonance imaging study. Journal of Neurophysiology, 92, 2428-2443.

Matelli, M., Luppino, G., and Rizzolatti, G. (1985). Patterns of cytochrome oxidase activity in the frontal agranular cortex of the macaque monkey. Behavioural Brain Research, 18, 125-136.

Merzenich, M. and Brugge, J.F. (1973). Representation of the cochlear partition on the superior temporal plane of the macaque monkey. Brain Research, 50, 275-296.

Mesulam, M.M. and Mufson, E.J. (1984). The insula of Reil in man and monkey. In A. Peters and E.G. Jones (Eds.), Cerebral Cortex. Vol. 4: Association and Auditory Cortices (pp. 179-226). New York: Plenum Press.

Meynert, T. (1867). Der Bau der Grosshirnrinde und seine ortlichen Verschiedenheiten, nebst einem pathologisch–anatomischen Corollarium. Vierteljahresschrift für Psychiatrie, 1, 77-93 (also, 1, 198-217).

Meynert, T. (1885). A clinical treatise on diseases of the fore-brain based upon a study of its structure, functions, and nutrition. Part I. The anatomy, physiology, and chemistry of the brain. New York: G.P. Putnam's Sons.

Mohr, J.P. (1976). Broca's area and Broca's aphasia. In H. Whitaker and H.A. Whitaker (Eds.), Studies in Neurolinguistics (Vol. 1, pp. 201-233). New York: Academic Press.

Mohr, J.P., Pessin, M.S., Finkelstein, S., Funkenstein, H.H., Duncan, G.W., and Davis, K.R. (1978). Broca aphasia: Pathologic and clinical. Neurology, 28, 311-324.

Morel, A., Garraghty, P.E., and Kaas, J.H. (1993). Tonotopic organization, architectonic fields, and connections of auditory cortex in macaque monkeys. Journal of Comparative Neurology, 335, 437-459.

Morosan, P., Rademacher, J., Schleicher, A., Amunts, K., Schormann, T., and Zilles, K. (2001). Human primary auditory cortex: cytoarchitectonic subdivisions and mapping into a spatial reference system. NeuroImage, 13, 684-701.

Nachev, P., Kennard, C., and Husain, M. (2008).

Functional role of the supplementary and pre-supplementary motor areas. Nature Reviews Neuroscience, 9, 856-869.

Nadeau, S.E., Rothi, L.J.G., and Crosson, B. (Eds.). (2000). Aphasia and Language: Theory to Practice. New York: Guilford Press.

Nagao, M., Takeda, K., Komori, T., Isozaki, E., and Hirai, S. (1999). Apraxia of speech associated with an infarct in the precentral gyrus of the insula. Neuroradiology, 41, 356-357.

Nota, Y. and Honda, K. (2004). Brain regions involved in motor control of speech. Acoustical Science and Technology, 25, 286-289.

Ojemann, G.A. (1979). Individual variability in cortical localization of language. Journal of Neurosurgery, 50, 164-169.

Ojemann, G.A. (1983). Brain organization for language from the perspective of electrical stimulation mapping. Behavioral Brain Research, 2, 189-230.

Ojemann, G. (1992). Localization of language in frontal cortex. Advances in Neurology, 57, 361-368.

Ojemann, G.A. and Whitaker, H.A. (1978). Language localization and variability. Brain and Language, 6, 239-260.

Ojemann, G., Ojemann, J., Lettich, E., and Berger, M. (1989). Cortical language localization in left, dominant hemisphere. An electrical stimulation mapping investigation in 117 patients. Journal of Neurosurgery, 71, 316-326.

Pandya, D.N. and Sanides, F. (1973). Architectonic parcellation of the temporal operculum in rhesus monkey and its projection pattern. Zeitschrift für Anatomie und Entwicklungsgeschichte, 139, 127-161.

Pandya, D.N. and Seltzer, B. (1982). Intrinsic connections and architectonics of posterior parietal cortex in the rhesus monkey. Journal of Comparative Neurology, 204, 196-210.

Parker, G.J., Luzzi, S., Alexander, D.C., Wheeler Kingshott, C.A., Ciccarelli, O., and Lambor Ralph, M.A. (2005). Lateralization of ventra and dorsal auditory-language pathways in th human brain. NeuroImage, 24, 656-666.

Paulesu, E., Frith, C., and Frackowiak, R.S. (1993) The neural correlates of the verbal componen in working memory. Nature, 362, 342-345.

Paus, T., Petrides, M., Evans, A.C., and Meyer, E. (1993). Role of the human anterior cingulate cortex in the control of oculomotor, manual and speech responses: a positron emission tomography study. Journal of Neurophysiology, 70, 453-469.

Penfield, W. and Boldrey, E. (1937). Somatic motor and sensory representation in cerebral cortex of man as studied by electrical stimulation. Brain, 60, 389-443.

Penfield, W. and Rasmussen, T. (1950). The Cerebral Cortex of Man—A Clinical Study of Localization of Function. New York: Macmillan.

Penfield, W. and Roberts, L. (1959). Speech and Brain Mechanisms. Princeton, NJ: Princeton University Press.

Penfield, W. and Welch, K. (1951). The supplementary motor area of the cerebral cortex. A clinical and experimental study. Archives of Neurology and Psychiatry, 66, 289-317.

Petkov, C.I., Kayser, C., Augath, M., and Logothetis, N.K. (2006). Functional imaging reveals numerous fields in the monkey auditory cortex. Public Library of Science Biology, 4, e215.

Petrides, M. (2002). The mid-ventrolateral prefrontal cortex and active mnemonic retrieval. Neurobiology of Learning and Memory, 78, 528-538.

Petrides, M. (2006). Broca's area in the human and

the non-human primate brain. In A. Amunts and Y. Grodzinsky (Eds.), Broca's Region (pp. 31-46). Oxford: Oxford University Press.

Petrides, M. (2012). The Human Cerebral Cortex. An MRI Atlas of the Sulci and Gyri in MNI Stereotaxic Space. New York: Academic Press.

Petrides, M. (2013). The mid-dorsolateral pre-fronto-parietal network and the epoptic process. In D.T. Stuss and R.T. Knight (Eds.), Principles of Frontal Lobe Function (2nd ed., pp. 79-89). New York: Oxford University Press.

Petrides, M. and Pandya, D.N. (1984). Projections to the frontal cortex from the posterior parietal region in the rhesus monkey. Journal of Comparative Neurology, 228, 105-116.

Petrides, M. and Pandya, D.N. (1988). Association fiber pathways to the frontal cortex from the superior temporal region in the rhesus monkey. Journal of Comparative Neurology, 273, 52-66.

Petrides, M. and Pandya, D.N. (1994). Comparative architectonic analysis of the human and the macaque frontal cortex. In F. Boller and J. Grafman (Eds.), Handbook of Neuropsychology (Vol. 9, pp. 17-58). Amsterdam: Elsevier.

Petrides, M. and Pandya, D.N. (2002). Comparative architectonic analysis of the human and the macaque ventrolateral prefrontal cortex and corticocortical connection patterns in the monkey. European Journal of Neuroscience, 16, 291-310.

Petrides, M. and Pandya, D.N. (2006). Efferent association pathways originating in the caudal prefrontal cortex in the macaque monkey. Journal of Comparative Neurology, 498, 227-251.

Petrides, M. and Pandya, D.N. (2007). Efferent association pathways from the rostral prefrontal cortex in the macaque monkey. Journal of Neuroscience, 27, 11573-11586.

Petrides, M. and Pandya, D.N. (2009). Distinct parietal and temporal pathways to the homologues of Broca's area in the monkey. Public Library of Science Biology, 7, e1000170.

Petrides, M. and Pandya, D.N. (2012). The frontal cortex. In J.K. Mai and G. Paxinos (Eds.), The Human Nervous System (3rd ed., pp. 988-1011). San Diego: Elsevier Academic Press.

Petrides, M., Alivisatos, B., and Evans, A.C. (1995). Functional activation of the human ventrolateral frontal cortex during mnemonic retrieval of verbal information. Proceedings of the National Academy of Sciences USA, 92, 5803-5807.

Petrides, M., Cadoret, G., and Mackey, S. (2005). Orofacial somatomotor responses in the macaque monkey homologue of Broca's area. Nature, 435,1235-1238.

Petrides, M., Alivisatos, B., Meyer, E., and Evans, A.C. (1993). Functional activation of the human frontal cortex during the performance of verbal working memory tasks. Proceedings of the National Academy of Sciences USA, 90, 878-882.

Poremba, A., Saunders, R.C., Sokoloff, L., Crane, A., Cool, M., and Mishkin, M. (2003). Functional mapping of the primate auditory system. Science, 299, 568-572.

Price, C.J. (2000). The anatomy of language: contributions from functional neuroimaging. Journal of Anatomy, 197, 335-359.

Price, C.J. (2010). The anatomy of language: a review of 100 fMRI studies published in 2009. Annals of the New York Academy of Sciences, 1191, 62-88.

Price, C.J. and Devlin, J.T. (2003). The myth of the visual word form area. NeuroImage, 19, 473-481.

Pulvermüller, F. (2002). The Neuroscience of Language: On Brain Circuits of Words and

Serial Order. Cambridge: Cambridge University Press.

Rademacher, J., Caviness, V.S.Jr., Steinmetz, H., and Galaburda, A.M. (1993). Topographical variation of the human primary cortices: Implications for neuroimaging, brain mapping, and neurobiology. Cerebral Cortex, 3, 313-329.

Rademacher, J., Morosan, P., Schormann, T., Schleicher, A., Werner, C., Freund, H.J., and Zilles, K. (2001). Probabilistic mapping and volume measurement of human primary auditory cortex. NeuroImage, 13, 669-683.

Rasmussen, T. and Milner, B. (1975). Clinical and surgical studies of the cerebral speech areas in man. In K.J. Zulch, O. Creutzfeldt, and G.C. Galbraith (Eds.), Cerebral Localization (pp. 238-257). New York: Springer-Verlag.

Rauschecker, J.P. and Tian, B. (2000). Mechanisms and streams for processing of "what" and "where" in auditory cortex. Proceedings of the National Academy of Sciences USA, 97, 11800-11806.

Riecker, A., Ackermann, H., Wildgruber, D., Dogil, G., and Grodd, W. (2000). Opposite hemispheric lateralization effects during speaking and singing at motor cortex, insula and cerebellum. NeuroReport, 11, 1997-2000.

Rilling, J.K., Glasser, M.F., Preuss, T.M., Ma, X., Zhao, T., Hu, X., and Behrens, T.E. (2008). The evolution of the arcuate fasciculus revealed with comparative DTI. Nature Neuroscience, 11, 426-428.

Rizzolatti, G. and Luppino, G. (2001). The cortical motor system. Neuron, 31, 889-901.

Romanski, L.M., Tian, B., Fritz, J., Mishkin, M., Goldman-Rakic, P.S., and Rauschecker, J.P. (1999). Dual streams of auditory afferents target multiple domains in the primate prefrontal cortex. Nature Neuroscience, 2, 1131-1136.

Rostomily, R.C., Berger, M.S., Ojemann, G.A., and Lettich, E. (1991). Postoperative deficits and functional recovery following removal of tumors involving the dominant hemisphere supplementary motor area. Journal of Neurosurgery, 75, 62-68.

Rushworth, M.F., Behrens, T.E., and Johansen-Berg, H. (2006). Connection patterns distinguish 3 regions of human parietal cortex. Cerebral Cortex, 16, 1418-1430.

Sarkissov, S.A., Filimonoff, I.N., Kononowa, E.P., Preobraschenskaja, I.S., and Kukuew, L.A. (1955). Atlas of the Cytoarchitectonics of the Human Cerebral Cortex. Moscow: Medgiz.

Saur, D., Kreher, B.W., Schnell, S., Kummerer, D., Kellmeyer, P., Vry, M.S., Umarova, R., Musso, M., Glauche, V., Abel, S., Huber, W., Rijntjes, M., Hennig, J., and Weiller, C. (2008). Ventral and dorsal pathways for language. Proceedings of the National Academy of Sciences USA, 105, 18035-18040.

Saygin, A.P., Dick, F., Wilson, S.M., Dronkers, N.F., and Bates, E. (2003). Neural resources for processing language and environmental sounds: evidence from aphasia. Brain, 126, 928-945.

Segal, E. and Petrides, M. (2012a). The morphology and variability of the caudal rami of the superior temporal sulcus. European Journal of Neuroscience, 36, 2035-2053.

Segal, E. and Petrides, M. (2012b). The anterior superior parietal lobule and its interactions with language and motor areas during writing. European Journal of Neuroscience, 35, 309-322.

Segal, E. and Petrides, M. (2013). Functional activation during reading in relation to the sulci of the angular gyrus region. European Journal of Neuroscience, in press.

Seltzer, B. and Pandya, D.N. (1978). Afferent cortical connections and architectonics of the superior temporal sulcus and surrounding cortex in the rhesus monkey. Brain Research, 149, 1-24.

Seltzer, B. and Pandya, D.N. (1984). Further

observations on parieto-temporal connections in the rhesus monkey. Experimental Brain Research, 55, 301-312.

Shellshear, J.L. (1927). The evolution of the parallel sulcus. Journal of Anatomy, 61, 267-278.

Signoret, J.L., Castaigne, P., Lhermitte, F., Abelanet, R., and Lavorel, P. (1984). Rediscovery of Leborgne's brain: Anatomical description with CT scan. Brain and Language, 22, 303-319.

Sotiropoulos, S.N., Jbabdi, S., Xu, J., Andersson, J.L., Moeller, S., Auerbach, E.J., Glasser, M.F., Feinberg, D., Lenglet, C., Van Essen, D.C., Ugurbil, K., Behrens, T.E.J., and Yacoub, E.S. (2013). The human connectome project: Advances in diffusion MRI acquisition and preprocessing. In Proceedings ISMRM 21st Scientific Meeting (pp. 52). Salt Lake City.

Stromswold, K., Caplan, D., Alpert, N., and Rauch, S. (1996). Localization of syntactic comprehension by positron emission tomography. Brain and Language, 52, 452-473.

Sweet, R.A., Dorph-Petersen, K.A, and Lewis, D.A. (2005). Mapping auditory core, lateral belt, and parabelt cortices in the human superior temporal gyrus. Journal of Comparative Neurology, 491, 270-289.

Talairach, J. and Tournoux, P. (1988). Co-planar Stereotactic Atlas of the Human Brain: 3 Dimentional Proportional System: An Approach to Cerebral Imaging. New York: Thieme Medical Publishers.

Thompson-Schill, S.L., D'Esposito, M., Aguirre, G.K., and Farah, M.J. (1997). Role of left inferior prefrontal cortex in retrieval of semantic knowledge: a re-evaluation. Proceedings of the National Academy of Sciences USA, 94, 14792-14797.

Tomaiuolo, F., MacDonald, J.D., Caramanos, Z., Posner, G., Chiavaras, M., Evans, A.C., and Petrides, M. (1999). Morphology, morphometry and probability mapping of the pars opercularis of the inferior frontal gyrus: An in vivo MRI analysis. European Journal of Neuroscience, 11, 3033-3046.

Türe, U., Yaşargil, D.C.H., Al-Mefty, O., and Yaşargil, G. (1999). Topographic anatomy of the insular region. Journal of Neurosurgery, 90, 720-733.

Ungerleider, L.G. and Mishkin, M. (1982). Two cortical visual systems. In D.J. Ingle, M.A. Goodale, and R.J.W. Mansfield (Eds.), Analysis of Visual Behavior (pp. 549-586). Cambridge, MA: MIT Press.

Upadhyay, J., Hallock, K., Ducros, M., Kim, D.S., and Ronen, I. (2008). Diffusion tensor spectroscopy and imaging of the arcuate fasciculus. NeuroImage, 39, 1-9.

Vernooij, M.W., Smits, M., Wielopolski, P.A., Houston, G.C., Krestin, G.P., and van der Lugt, A. (2007). Fiber density asymmetry of the arcuate fasciculus in relation to functional hemispheric language lateralization in both right- and left-handed healthy subjects: a combined fMRI and DTI study. NeuroImage, 35, 1064-1076.

Walker, A.E. (1940). A cytoarchitectural study of the prefrontal area of the macaque monkey. Journal of Comparative Neurology, 73, 59-86.

Wernicke, C. (1874). Der aphasische Symptomencomplex. Eine psychologische Studie auf anatomischer Basis. Breslau: M. Cohn and Weigert.

Wernicke, C. (1881). Lehrbuch der Gehirnkrankheiten: für Aerzte und Studierende. Kassel: Theodor Fischer.

Willmes, K. and Poeck, K. (1993). To what extent can aphasic syndromes be localized? Brain, 116, 1527-1540.

Wise, R.J.S., Greene, J., Buechel, C., and Scott, S.K. (1999). Brain regions involved in articulation. Lancet, 353, 1057-1061.

Woolsey, C.N., Erickson, T.C., and Gilson, W.E. (1979). Localization in somatic sensory and motor areas of human cerebral cortex as determined by direct recording of evoked potentials and electrical stimulation. Journal of Neurosurgery, 51, 476-506.

Zhu, Z., Hagoort, P., Zhang, J.X., Feng, G., Chen, H-C., Bastiaansen, M.C.M., and Wang, S. (2012). The anterior left inferior front gyrus contributes to semantic unification NeuroImage, 60, 2230-2237.

Zlatkina, V. and Petrides, M. (2010). Morphologica patterns of the postcentral sulcus in the huma brain. Journal of Comparative Neurology, 518 3701-3724.

Printed and bound by CPI Group (UK) Ltd, Croydon, CR0 4YY

08/05/2025

01865034-0004